MW01010386

HUMAN FOSSIL RECORD *and* CLASSIFICATION

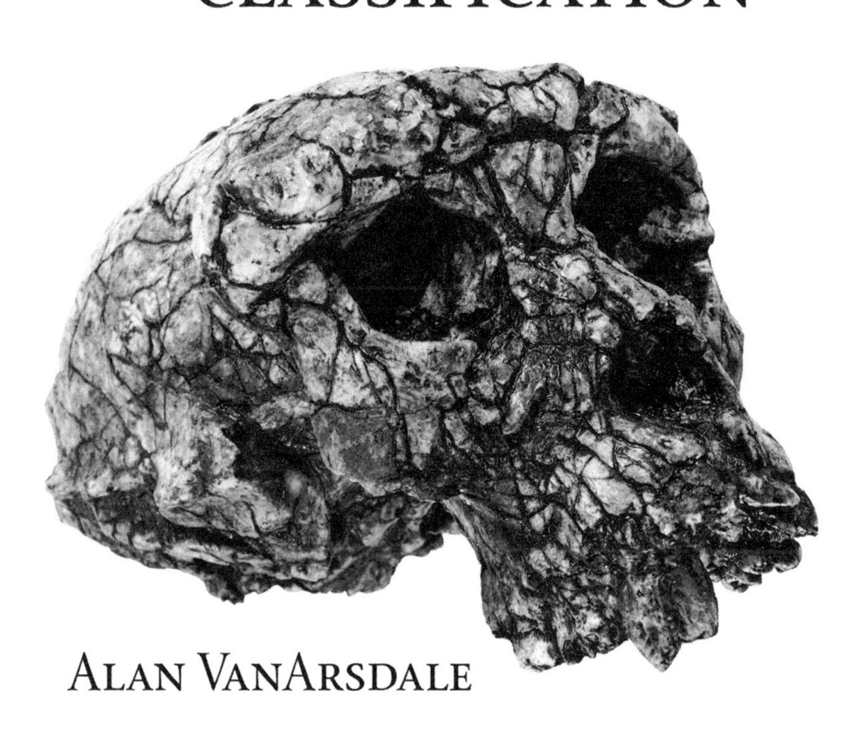

Alan VanArsdale

ISBN 978-1-09830-456-0

CONTENTS

INTRODUCTION

This book was written for use by the interested public as well as professionals in related fields. At the end of the book is a glossary of words and terms used in the book defined as the author understands them. The intention of this book is to serve as a modernized synthesis of existing work and finds in summary. From the authors perspective the most likely and parsimonious views are presented, there are other valid views.

Especially in the last 25 years, many new fossils have come to light in Asia, and much data and interpretation of data has come forward from genetics as well as other fields. Severely challenging some prevailing opinions among professionals in the 1990's, while supporting the findings of other sometimes obscure professionals and amateurs. With paleoanthropology and archaic human genetics in constant states of revisions for the last 15 years. Science tends to build upwards from bases, until those bases are widely agreed to have been disproven. Given the short length of this book, relative to the broad and deep subjects, here this process is somewhat reversed. Building often on the most recent findings in a top down process.

So much has been written about these topics, by professionals and amateurs alike, few stones remain unturned in terms of speculation and interpretations. Nobody has such a broad knowledge of everything written about human paleontology (paleoanthropology the study of human and human related fossils) / systematics (how the fossils are related to each other) as to know always if they are repeating something which has been written some place. Unreferenced statements in this book may be the original work of the author, work of other authors which this author is unaware of, or the work of other authors this author is aware of, but which has not located among the thousands of relevant works the author has read.

Social media has brought people together and created venues in which researchers can rise, or fall, in public and peer perception, outside of the peer reviewed literature and press. Journal writers are not paid for their submissions, except in reputation. With increasing use of social media to research and interact, and less use of great libraries of physical books, issues about public free access to the literature without travel to great libraries have increased in importance. Concepts about what constitutes peer review, or scientific publishing, are changing. The author had no help with the actual book. And much help from many people over the many years leading up to this book.

CHAPTER 1

Interpretation of the Human Fossil Record

Fossil morphology external and internal is absolute truth. If there is no incorrect or composite reconstruction of fossils. Damaging fossils such as by poor preparation or removal, does not add false information it removes information. Where we all often go wrong is in the interpretation of the absolute reality presented to us by fossils. In human genetics this problem of misinterpretation of data is greater than in paleoanthropology. Genes are also fossils, evidence of past life passed down by dead people to us. Human geneticists have a very large data base for people of the last few thousand years, all fully modern humans (AMHS anatomically modern Homo sapiens). And a very small data base for ancient people who were not fully modern. Only two genomic groups, neandertals and denisovans. Both relatively low population extreme cold adapted populations in geographical and kinship relationship contact with each other. Neither of which groups were thought in the 1990's to be ancestral to people today, with exception of many Chinese professionals a few Western professionals and some amateur researchers including this author. Along with a few odd genetic relic ghosts of archaic humans or Homo erectus, which can't be reliably attributed to any geographic or known fossil origin.

The press and public usually are behind professionals and their amateur peers in understanding paleoanthropology. A fact often lamented by professionals both in print and social media. This book attempts to correct this situation some. Linear evolution is not widely accepted today by professional paleoanthropologists or evolutionary biologists yet remains engrained in the public mind. Embodied by charts showing chimpanzee like apes rising in a linear progression of ape men to become fully modern. 19th century Victorian grade typology is well entrenched in Western paleoanthropology until today, it has not served the field well. The fossil record more and more, and especially genetics, suggests a reality more like spectrums of morphologies gently grading into one another in small clinal steps than strongly distinct morphotypes clearly defined by a single holotype. Field paleoanthropologists are not looking for missing links, they are looking for fragments of spectrums of diversity. Human history is not a chain through time, it is more like a complex braided stream with channels coming in and out.

Franz Weidenreich was a medical Dr anatomist and physical anthropologist. In his 1946 book "Apes Giants and Man", he wrote human populations same as today, have always mixed. At the time, most of his peers were typologists and considered that human races began as pure races without mixing, and sometimes later mixed with other races. This latter view has especially suffered in the last few years under the weight of genetic evidence. When neandertal mt DNA were first discovered, it was published mt DNA evidenced no neandertal ancestry in people living today. David Reich and his team, against their own training and expectations, later found compelling evidence humans today have derived neandertal genes outside of Africa. Dr Loring Brace and Dr Adcock disputed the initial results in published papers before human geneticists reversed their opinion about neandertal genomic material not having contributed to extant humans at all.

Since the late 1970's most Chinese paleoanthropologists and Loring Brace (along with a few others outside of Asia), a Harvard PhD studying many human remains, have rejected typological interpretations of human classification. They rejected notions that isolated fossils can adequately represent biological species of humans. Rejecting that populations of humans have often been genetically isolated from other populations. And rejecting the full replacement of Eurasians by small African founding population(s), or the replacement hypothesis (ROoA) which is now nearly universally rejected without amendment. They also rejected convergence (independent development of traits) for sets of human character states is naturally more parsimonious (simpler so more likely correct), versus such trait sets being shared as introgressive traits (traits moving about by gene flow / interbreeding) or ancestral. The continuity hypothesis predicts that trait complexes seen in early humans, and in extant humans in the same regions, usually were passed down by inheritance, not convergence or similar independent adaptations to similar environments.

Gene pools often become fragmented into different pools (biological speciation). And rarely detectably flow back and forth in derived traits once true biological speciation has taken place. Normally when hybrids between species are fertile, they back breed into one or the other parent species and disappear in the gene pool morphologically. The process of biological speciation does not usually take place quickly with primates, early Paleocene mammals, or hominins (populations more closely related to us than chimpanzees). The concept of LCA (Last Common Ancestor), is easy to teach, leads to nice clear linear charts which get press and grants. However, the concept of LCA (a quick speciation in a relatively uniform population), is not generally valid for the human adaptive radiation. Normally the timing between when populations become distinct from each other morphologically in hominins, and great apes. to when they form biological species, is long. On the order of a half a million to four million years. A long and messy processes cladists dislike in favor of simple linear divergences in cladograms. Hominin reality involves a lot of sex between different populations usually yielding fertile offspring, for very long periods. Nothing like apes giving birth to Adam and Eve, with speciation in a century or two with a few apes and giants marrying humans, and the apes going merrily along their way as living fossils without change. Weidenreich in 1946 was already aware that humans (the genus Homo), is just one biological species. Irrespective of high morphological variation, though not nearly so high as in some early Paleocene mammal species.

The fossil record for hominins is not a lineage with a few missing links absent. It is like a large puzzle with many thousands of pieces, each piece being a population, often represented by just one or a few individuals. Not only are most of the pieces missing, there are very large geographic areas and time frames with no pieces at all on all continents including Africa. Trying to jam the pieces we have into a nearly linear progression like a column through time, leads to many incorrect perceptions. Lots of room needs to be left between the pieces to give a better feeling for how the big picture looked, with large areas completely empty. Genes which determine morphology, which is all we see in fossils when no biochemical remains are present from the person who made the remains, are only a tiny fraction of the whole genome, a few percent. Nor do geneticists know in most cases where the morphology genes are in the genome for known genomes, and they know almost nothing about morphology genes in archaic humans except neandertals and denisovans.

Denisovans are not a taxon, they are the first human grouping based upon genes alone. There are a few Denisovan teeth and bone fragments known from Denisova Cave in Siberia, a lower jaw identified by proteonics from Tibet, and some other fossils attributed as likely denisovans based upon morphology, especially Homo tsaichangensis another lower jaw. The origin of the Denisovan genes defining the group is unknown. This author suspects they come from cold adapted Asiatic Homo erectus, Meganthropus, unknown ancestral Gigantopithecus, or other unknown Asiatic populations. Paleoanthropology from fossils sees only the morphological part of the genome, and arguably evidence of some genes related to the brain feet and hands.

Often looking at whole genomic distances to classify hominins and great apes gives quite a different impression than the study of fossils. Genes do not usually move together they pass around in small packages or alone acted upon by natural selection to genomes not having the genes. When there are no fossil genomes, living taxa have altered over time considerably from ancestral populations so are seldom good models for ancestral forms. When groups of genes need to all be present to impart adaptive advantage, the groups are sometimes maintained together in a population by natural selection and individuals with the gene groups fragmented are selected out. Usually until that part of the genome in a population has few competing genes to dilute important complexes. As with Denisovan genes in people today which are usually clumped together when detected.

As genes move through time and space, they meet new conditions which sometimes disfavor them and create barriers to their passage. Genes can follow paths of opportunity, such as following deadly epidemics into temperate areas from tropical regions, leaving the rest of the parent genome behind. The Y chromosome (patrilinear) is only a small part of the whole genome, effecting male morphology to some degree, and often does not move with the rest of the genome over time and distance. The mt DNA (matrilinear) is only a tiny fraction of our whole genomes, and moving back in time, also usually has very little to do with whole human ancestry and morphology. Sometimes mt DNA is contaminated with male sperm DNA when the tail of the sperm does not fully degrade before entering the egg. Otherwise it is passed down matrilineages from mother to daughter.

We have no genes at all in us from most of our distant individual direct ancestors (people in our family tree). Taken out at random over time, or because they became disfavored under natural selection by changes in the natural environment, culture, or material culture. Most people who ever lived left no descendants today. There is no good reason to think that any population of Homo that ever lived, is not represented by in people today genetically. Purative selection and random walk are unlikely to entirely cleanse any group of genes from an ancestral population entirely from the gene pool, though usually they will remain undetected until now. Genes giving local advantages, or advantages to all people in the World so becoming universally shared, are as easily spread from a small population as a large one.

Lee Berger has been promoting the braided stream model of human evolution, as has this author on face book for four years. Tim White does not think there is any need for a new model. In the braided stream model through time the gene pool is like a braided stream. Genes flowing underground, so no longer expressed in the phenotype. Being lost through evaporation (negative selection). Genes are lost to the primary channel but are regained as they flow back in from side branches. Mutations and HGT (genes being brought from other species in retrovirus and other microorganisms), seen as rain falling into the braided stream from outside the river valley.

CHAPTER 2

Review of the Hominin Fossil Record

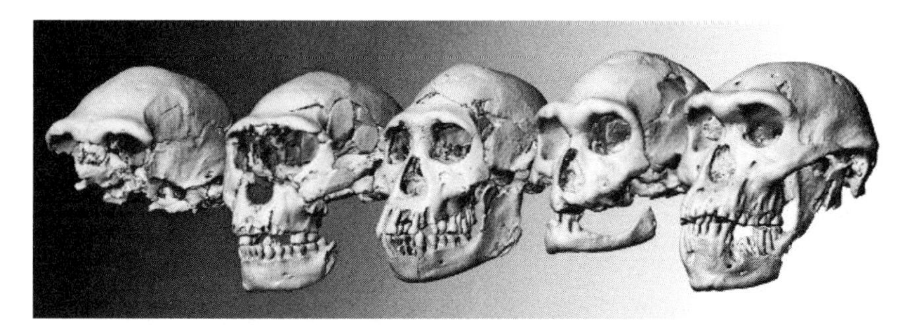

(Image credit: Image courtesy of M. Ponce de Leon and Ch. Zollikofer, University of Zurich, Switzerland). The five Dmansi skulls 1.8 million years ago.

PHOTO - Remarkable morphological diversity seen in the five skulls recovered from Dmansi Georgia of the most primitive / basal Homo erectus known. Found at Dmansi Republic of Georgia in Western Asia. Adam Van Arsdale has explained the diversity is in part due to sexual dimorphism and different ages upon death. The fourth is an elderly individual that would have required help to survive with no teeth so late in life before his death. If found at different locations in Africa in the 1960's multiple species would have been named from these skulls. Since these finds standards have become more difficult than before to name new species of hominins. In living humans Homo sapiens sapiens morphological diversity between all populations is much higher than seen in the single spot find sample from Dmansi. In hominins, basal Homo sapiens, Homo erectus and basal likely hominins all demonstrate higher morphological diversity earlier than later. Suggesting the more common pattern is for the highest diversity being at the beginning of human radiations, lowering over time as the case now, and increasing again before the next adaptive radiation / founding of new species.

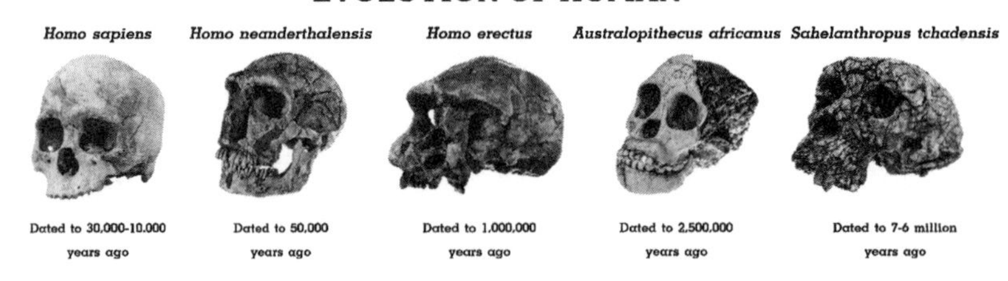

Chart credit – Puwdol Jaturawutthichai

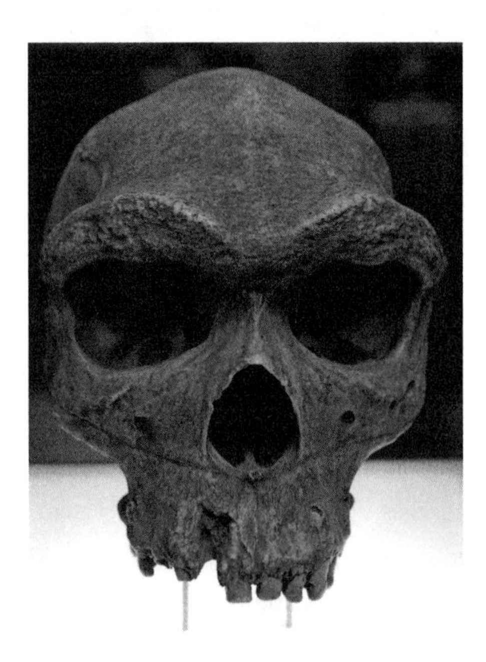

Photo Homo heidelbergensis

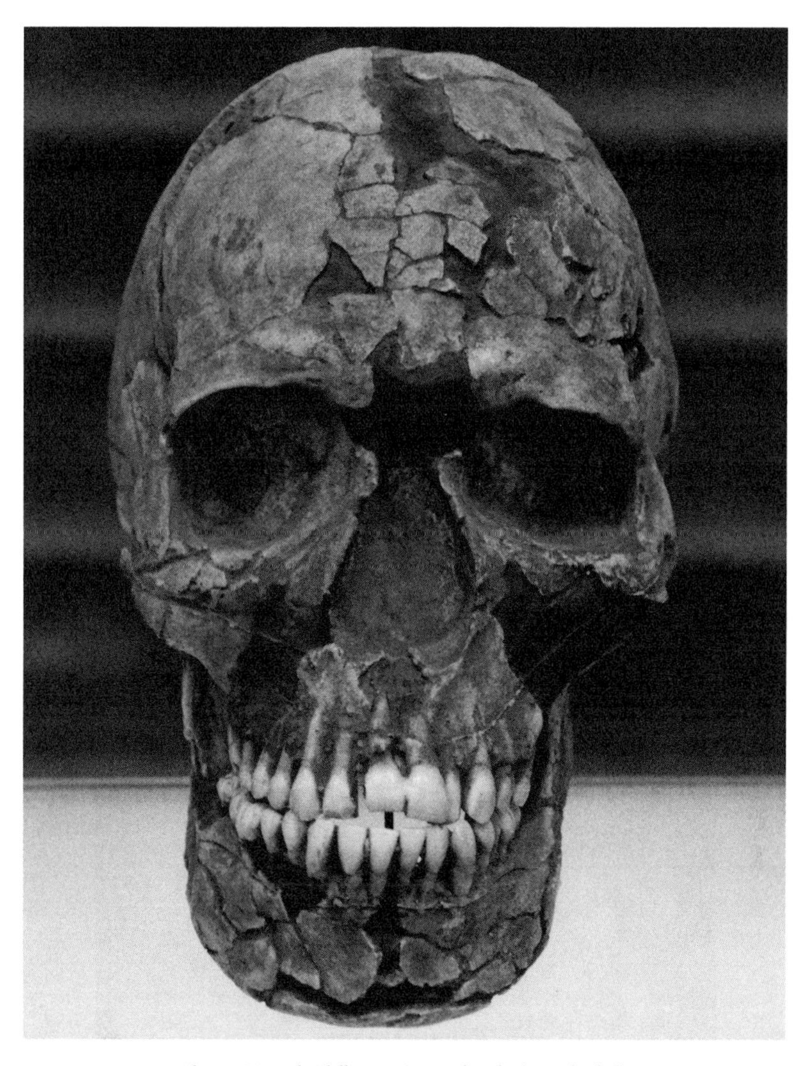

Photo – Homo heidelbergensis neandertalis (Neanderthal)

Photo - Sahelanthropus tchadenisis

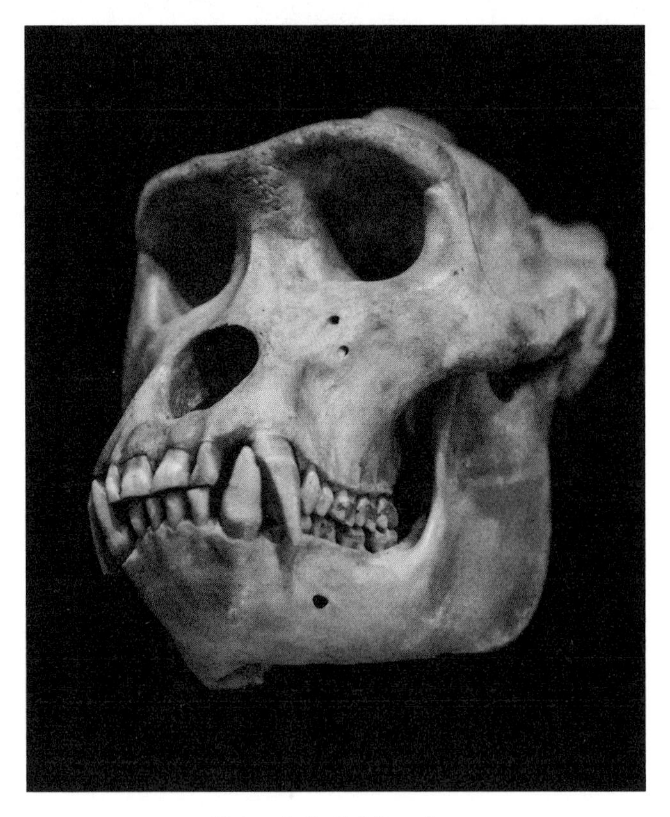

Photo male Gorilla fully modern

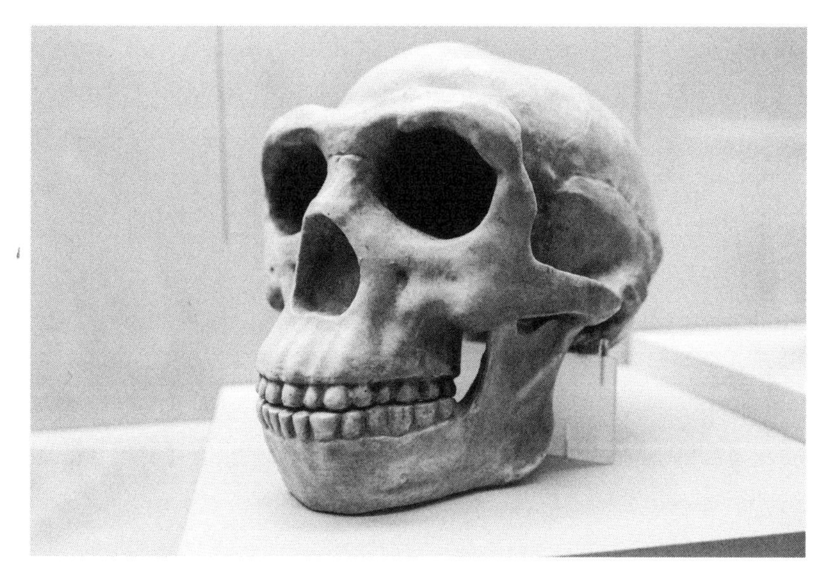

Photo – Homo erectus reconstruction

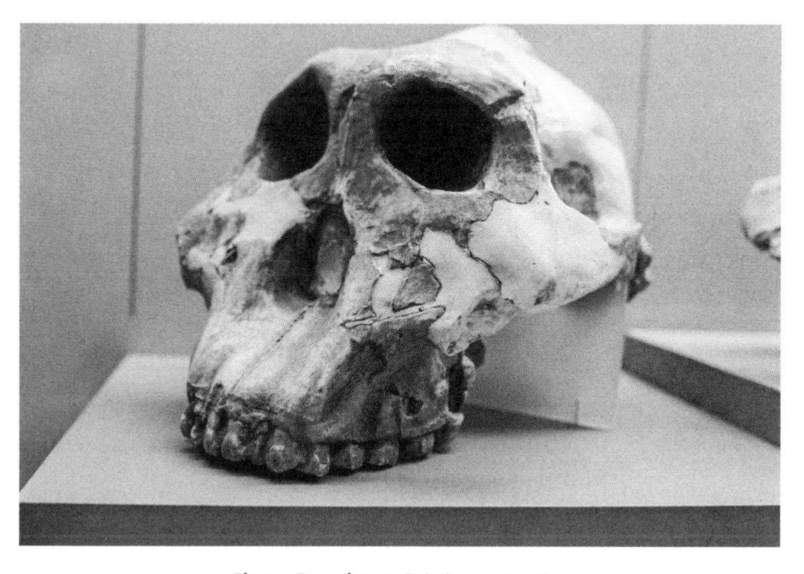

Photo – Paranthropus Boisei reconstruction

Discussion of photos. In the number of genes shared by chimpanzees, humans and Gorilla each pair of groups shares about the same number of genes to the exclusion of the third group. Chimpanzees are suggested to have speciated from humans later than Gorilla and are closer in whole genomic distances than living Gorilla to humans. However, fossil bipedal gorillinae very likely were at least as close to gracile bipedal Australopithecus genetically as primary quadrupedal (knuckle walking) ancestral chimpanzee populations were at the same times. The strong similarities to gorillinae of Paranthropus, Meganthropus, some Australopithcus, Sahelanthropus and even some Homo erectus to the exclusion of Homo sapiens is not always due to shared primitive traits from common ancestors. This similarity, as suggested by comparison of the above photos to Gorilla, are sometimes from introgression of derived gorillinae traits differentially (more into some less into others), into hominins (Gorilla related great apes and early hominins interbreeding). Today just as Asiatic great ape traits are more common and pronounced in East Asia than other places, gorillinae traits are more pronounced in Europe and Africa relative to other places. While most differences between different populations today come from archaic humans in people's respective regions, a small fraction of the differences reach back into great ape diversity. No below fully modern fossil gorillinae skulls are known, unless Sahelanthropus is one. So only recent Gorilla skulls are available for comparison, which are derived relative to ancestral gorillinae and thereby less hominin like.

The genus Australopithecus is the oldest widely agreed upon hominin, known only from Africa at about 4.2 to 1.4 million years ago. With at least seven widely recognized species named. Hominins might more correctly contain only two genera, Australopithecus and Homo, if the four likely basal hominin genera are considered great apes or placed in Australopithecus. In adaptive radiations especially, with mosaic evolution, where they are placed can be arbitrary great as in apes or hominins. Especially when mosaic mixtures of great ape and hominin traits are seen without the benefit of more than a few specimens unlike with Australopithecus. Four likely hominin genera have been named based upon fossils older than known Australopithecus, from oldest to youngest Graecopithecus, Ardipithecus, Orrorin and Sahelanthropus. There is no wide agreement on these early four likely to be hominin genera if they are hominins, great apes, or populations falling between or to the side of great apes and hominins. A population can appear to be between great apes and hominins because they diverged from great apes and hominins after their initial divergence, and before they speciated. Or because they represent populations which diverged from great apes or hominins after those groups speciated but retain characters from one or the other group not seen in later hominins or great apes. For example, early hominins, already speciated from great apes, could have had plesiomorphic traits from ancestral populations shared by both great apes and hominins, but not seen in later hominins. So "primitive" traits can seem to be diagnostic for either great apes or hominins but be present in both groups from common ancestral populations. Or intermediates can arise through introgressions between great apes and hominins after they diverged but had not speciated yet, sometimes with traits through introgressions from other apes or primates as well early in hominid adaptive radiations.

Fossils of hominins and likely hominins other than fully modern humans (AMHS) are normally very rare. Most places hominins lived no fossil mammals or hominins are found. Most places fossil mammals are found where hominins lived, no fossil hominins have been found yet. Usually where fossil hominins are found, for each hominin bone or tooth found, thousands of animal bones and teeth are found. Hominins from before Homo erectus times are represented by just one or a few individuals for each population / morphotype, and only fragments or a few bones and / or teeth from most of those. While all unique material is useful and important, it is often impossible, or nearly so, to make many conclusions about phylogenetic status (the relationships of the fossil(s)), to other hominins. Often records can be truncated (they are not found until long after they first existed). For examples the fossil record for hominins in Eurasia does not exist until at least 300,000 years after hominins are evidenced by tools and tool use in Eurasia. And there is substantial evidence there were humans in Sahul (a now partly submerged continent including Australia New Guinea and Tasmania), long before fossils are seen about 50,000 years ago. There are many geographic and temporal gaps in the fossil record, especially in areas which have heavy vegetation so no fossils can be found on the surface, have no terrestrial rocks on the surface of the correct ages, or had acidic soils which destroy bones and teeth before they became fossils (such as areas that were forested).

The oldest likely known hominin is a lower jaw found at Athens Greece in 1944 dated at 7.2 million years, Graecopithecus freybergi (Fuss et al 2017). With some isolated upper teeth from Macedonia which this author considers are likely from a different population which were great apes not hominins, though related to Graecopithecus. Graecopithecus lived in the Northern part of a savanna like environment which extended well into Africa, in which the mammals were typical of Africa for then and now. Continents are defined by geology and nations are political constructs. Early hominins were contained only by real obstacles such as high mountains and oceans, or environments they could not live in, not by geological or political barriers. They always existed over much wider ranges than they are known from as fossils. In the Miocene, when earliest hominins or Homo like great apes such as Lufengpithecus are first known, there were usually forested or other green corridors connecting Africa and Eurasia. The Arabo-Saharan deserts green on about a 23,000-year cycle (Jessica Tierny et al 2017), so do not present lasting obstacles to gene flow of hominins back and forth between Africa and Eurasia. The partial consensus age for the divergence of chimpanzees and humans, about 7 million years ago, is likely to be incorrect. Ven et al 2014 suggest a divergence of 11-17 million years ago in "Strong male bias drives germline mutation in chimpanzees". And in MGD genetic studies, which yield results much more in agreement with the fossil record than Western methods in genetics, the suggested divergence date much older as well. Nor was there an exact age for the divergence, it likely took millions of years, from initial morphological divergence to speciation, as is typical in primate divergences.

Graecopithecus shows considerable morphological affinities to both chimpanzees and Australopithecus, in addition to SE European fossil great apes. Based upon genetic studies, chimpanzees and hominins may have been populations of the same species 7.2 million years ago. Hominins at the time of Graecopithecus are likely to have interbred with ancestral chimpanzees enough to share many derived traits (traits not in their common ancestors) with chimpanzees. Graecopithecus appears to have been bipedal based upon the inside morphology of the anterior lower jaws, though the literature does not share this authors opinion that this morphology is useful in determining bipedalism. There is only one archaic chimpanzee fossil known, a chimpanzee "like" lower molar from Kenya dated at 12.5 million years (Pickford et al 2001), which are in this authors opinion is a Pan (chimpanzee) tooth. There were great apes living in SE Europe up until 7 million years ago, which were adapted to less forested conditions in which fruit was only seasonally available (Spassov et al 2012).

The second oldest likely hominin fossil is Sahelanthropus tchadensis first found in 2002, which is known only from a weathered skull, and an associated by not articulated weathered leg bone (unpublished but we have a photo), and some jaws and teeth found later at Djurab. All discovered by a team led by Michel Brunet in Chad Africa, questionably dated at about 7 million years (possibly younger so Orrorin may instead be the second oldest likely hominin known). John Hawks and Milford Wolpoff have cast doubt on Sahelanthropus as being either a biped or a hominin. This author finds arguments based upon the femur and skull convincing for Sahelanthropus to have had considerable bipedal abilities.

The human raised chimpanzee (Pan troglodytes) Oliver habitually walked bipedally. Bonobos are reported to spend about forty percent of their time on the ground bipedally. Now several species of fossil great apes, such as Lufengpithecus, are known to have been bipeds. Wood considered if we accept Sahelanthropus as a hominin, with a face as flat (low prognathy / jaws not jutting out much) as hominins one third the age of Sahelanthropus, then a long list of hominins must be excluded from the ancestry of Homo. Loring Brace did not consider prognathy very important in hominin classification, viewing the length of jaws as being more related to diet than phylogeny. Relatively low prognathy may be the primitive condition for great apes, as orangutans have low prognathy along with fully modern humans relative to Western (African and European) fossil and extant great apes and hominins.

The Sahelanthropus skull bears striking similarity to female gorillas. (Pickford and Senut 2001) concluded Sahelanthropus is more likely a female proto-gorilla than an hominin. With almost no fossil record for below modern grade great apes from Africa, it is not known how a female proto-gorilla should appear. Now that it is well evidenced bipedalism existed in great apes before hominins, and that great apes today can elect to walk bipedally comfortably in some individuals at least for all types of extant great apes, the presence of a bipedal proto-gorilla in Africa should not be surprising. A Sahelanthropus canine is worn on the tip, showing a honing process. As in hominins, and the great ape Gigantopithecus known from Asia and Germany and not otherwise known in great apes.

Orrorin tugenensis (Pickford et al 2001) is the third oldest likely hominin fossil known, dated to between 6.1 and 5.7 million years old. First found in Kenya Africa in 2000, the authors argued Orrorin demonstrates Australopithecus is not ancestral to Homo. (Almecija et al 2013) found Orrorin femurs (upper leg bones) to be intermediate between Miocene great apes and hominins, and not much like those of living great apes. Arguing some Miocene great apes are better models for hominin locomotion than any living great apes are. It has since become widely accepted that populations ancestral to both great apes and hominins did not look much like any great apes living today. Bones of Orrorin suggest bipedalism but without long term weight on just two legs sustainable as they are thin to suggest sustained bipedalism. Without any skulls it is difficult to decide how Orrorin is related to other hominins and great apes.

Going back to Greece, for the fourth oldest likely hominin fossils (Gierlinske et al 2017), from the Island of Crete in Greece SE Europe. Which at the time of the fossils was connected to the Asiatic Mainland and again a part of an African mammalian fauna. Reported likely hominin track ways in a marine (beach / bay) sediment reliably dated at 5.7 million years. The tracks are very Homo like in appearance, so for this reason alone generally rejected by Western paleoanthropologists as being hominin tracks (too old for Homo they think), with the invocation of convergence. Despite it is very unlikely any entirely unknown group of primates ever developed such a strikingly Homo like foot independently. This author identifies these tracks as Homo ?naledi though the original authors, expert ichnopaleontologists (paleontologists of animal tracks burrows etc.), but not experts on hominin feet, left it up to others to identify the tracks (which this author has done).

At this time the trending opinion among professionals is that Homo naledi and Homo floresiensis never had larger than ape sized brains, and that Australopithecus is not ancestral to Homo. If we accept the fossil record for Homo is badly truncated (except for these tracks), it becomes easy to accept they are Homo tracks just as they very much appear to be. In the fossil track ways from Crete one individual is larger than then the rest, suggesting a dominant male and Homo naledi level sexual dimorphism. The arches of the feet are well developed as in Homo but not in other types of known hominin tracks and feet. Track morphology and spacing suggests sprinters and able walkers but with mechanics and ligament strengths different than for any known hominin, though still much closer to Homo than any other taxa. With the forefoot bearing most of the weight and striking the ground first as is common in children now when sprinting. No tracks are otherwise attributed to Homo naledi though a nearly complete foot is known.

Ardipithecus ramidus (Tim White et al 1995) with a temporal range of 5.8 to 4.4 million years, was first discovered by a Native African paleoanthropologist as part of the Tim White team in the Afar depression of

Africa. First named as Australopithecus ramidus and a year later reassigned to a new genus Ardipithecus. Since then they have discovered and named a second species in the same genus also from the Afar of Ethiopia. Ardipithecus suggests bipedal ability with at the same time considerable tree climbing ability, though less so than chimpanzees. It had been thought great apes could be distinguished by relatively distally squared pallates (relatively squared at the front upper jaws), relatively high sexual dimorphism (males compared to females larger than for hominins, some Australopithecus push the line for sexual dimorphism in hominins), relatively larger canines, and the hallux (big toe) pointing out to the side of the foot. Ardipithecus has the big toe pointing out to the side and some other great ape characteristics, along with many hominin characteristics, and ape sized brains smaller than a chimpanzee brain today. Tim White proposed Ardipithecus was ancestral to Homo and has since backed off from that opinion. Some consider Ardipithecus rules out known Australopithecus as ancestral to Homo, because Ardipithecus shares some traits with Homo and Australopithecus which had been thought to have first appeared in Australopithecus and from there passed down to Homo. Ardipithecus lacked knuckle walking hand / wrist morphology suggesting at least some of the ancestral populations of both chimpanzees and humans were not knuckle walkers. Opinions have been mixed on this (Richmond et al 2000) presented morphological evidence that hominins evolved from knuckle walkers.

Australopithecus is the oldest primate genus which is almost undisputed as being a hominin. William H. Kimbel et al published a paper in 2016 "From Australopithecus to Homo the transition that wasn't". Australopithecus appears to be a hominin sister group of Homo, not ancestral to Homo as had been thought. Australopithecus however shows many signs of introgression both from African knuckle walking apes and Homo, it is not known at what point in time, if any, Australopithecus speciated from Homo or Paranthropus. Some Australopithecus were knuckle walkers with knuckle walker morphology in the hands and wrists. Which more likely is from introgression from African or other great apes than any indication they were great apes. Or that the knuckle walking trait was in some Australopithecus since the time they were great apes (plesiomorphic). Showing photos of Gorilla, Pan (chimpanzee), and some Australopithecus skulls together shows they are much more similar over all morphologically than they are to any known Homo. Orangutans also knuckle walk, as amply shown in films available in social media. Australopithecus vocalizations may have been more chimpanzee than human like, as their hyoid bone when known are structurally closer to those of chimps than humans (Steele et al 2013).

The type specimen of about 2.5 million years for Australopithecus was found in 1924 in South Africa, commonly called the Taung child an infant of three years age at death. Which was first described by Raymond Dart in 1925. Leakey after years of struggle began finding hominin fossils in East Africa, before that nearly all known hominin fossils other than fully modern ones were from Eurasia and Southern Africa. Fully modern humans relative to all other known hominins and great apes have more neotonous traits (they look more like children as adults). There are no known juvenile Miocene ape craniums except for Lufengpithecus from China, which look remarkably like the adult Homo floresiensis skull. Lufengpithecus milk teeth are strikingly like the Taung child milk teeth in dental morphology. Infant orangutans have high domed, well rounded craniums much like fully modern human adults and have much shorter jaws (are much less prognathous) than Taung child. The fact that young hominins and great apes in general tend to look like fully modern human adults gave the false impression that the Taung child was closely related to fully modern humans.

The genus Paranthropus is assigned by most authors for the last thirty years to the genus Australopithecus. Paranthropines are by most thought to be derived from Australopithecus. Paranthropines were robust (big strong and heavy boned with massive molars) hominins who appeared rather great ape like, filling great ape niches, and known much later than

Australopithecus. Paranthropus had sagittal crests, ridges of bone along the tops of their heads to attach strong chewing muscles to, a trait in common with male Gorilla and some "Meganthropus" from Indonesia and known in large chimpanzees. Paranthropus is known from about 2.6 to .6 million years ago. Paranthropines generally became larger and with bigger teeth over time, the last ones known being the most ape like. An example of convergent evolution or introgression from great apes or ape like basal hominins to paranthropines over time. The archaeological record suggests Paranthropines made and used Oldowan grade stone tools. In the paper "Paranthropus boisei: Fifty Years of Evidence and Analysis" Bernard A. Wood and Paul J. Constantino 2007 Paranthropus is proposed as a valid genus. Paranthropus can be identified by a single molar from all other hominins and great apes. Distinguishing Homo from Australopithecus can sometimes be uncertain even with a skull. Louis Leakey 1966 proposed that the hominin groups he had found up until 1966 evolved from older unknown groups, not from each other including Homo and Australopithecus.

All hominin genera in commonly used with diagnostic material known have been considered in this chapter. Meganthropus has been reassigned to Homo erectus in the professional literature, in the common belief that all Eurasiatic hominins belong to the genus Homo and are evolved from Homo erectus or above grade hominins (since 2017 this is no longer the case for Homo floresiensis). Many amateur researchers have always considered the genus Meganthropus to be a valid taxon. Meganthropus, known only from the Pleistocene of Indonesia, grades into Javanese Homo erectus. Suggesting Meganthropus evolved from Homo erectus. Or Meganthropus could be a type of hominin which never was Homo erectus that was able to interbreed with Homo erectus upon contact. Such as a paranthropine or basally divergent Gigantopithecus.

One of the reasons given to move Meganthropus to Homo erectus, was stone tool association for Meganthropus, in the belief that only members of the genus Homo made and used stone tools. Now capuchin monkeys are known to make Homo grade flakes from stone cores. Chimpanzees (Pan troglodytes), are known to use stone tools. There is Oldowan grade stone tool association known for a very late occurrence of the bipedal great ape genus Lufengpithecus, at 200 thousand years. And there is compelling stone tool association for Paranthropus and Australopithecus. Some Homo erectus paleojavicanus (Meganthropus), are certain hominins. Others are morphological great apes. Kenyanthropus from about 3.5 million years ago is represented by a skull composed of many fragments skillfully put together. The position of Kenyanthropus relative to other hominins is uncertain, possibly it merits generic status or is an Australopithecus or Homo.

Clement Zanolli et al 2019, found that some of the more robust dental and jaw fragments originally assigned to Meganthropus, are instead from unknown great ape(s). Great apes not closely related to orangutans or Gigantopithecus, the only great apes previously known from Indonesia. Nothing is known about Asiatic great ape genomes except for fully modern orangutans. Australopithecus may have given rise to Paranthropus, possibly with introgression or HGT from African great apes at some point. Paranthropus is great ape like and Meganthropus like, through convergence and / or introgression, or by close relationship. Homo tsaichangensis, probably a Denisovan, suggests unlikely levels of convergence between great apes and Homo in Asia, or introgression from great apes Gigantopithecus or Paranthropus at some point in their history independent of known African Homo. If the introgression is from Paranthropus, then the introgressing unknown paranthropines suggest introgression from Gigantopithecus to a level not seen in Africa in Paranthropus. Gigantopithecus is known from Germany and is not known from Africa and was widely thought to be a hominin until at least 1973, for good reasons.

Gibbons and their relatives (Hylobatidae) are known only from Asia now and in the fossil record and among apes are more closely related to great apes and humans than any other known type of ape. Most other fossil apes are more likely to be large monkeys than closely related to great apes. With no extinct Asiatic great ape genomes known, and details about the timing of their divergences unknown, the timing of Asiatic great ape speciation from the genus Homo is also unknown. Grehan and Schwartz 2011, as part of ongoing very detailed cladistic and other analysis, have found that Orangutans form a morphological clade with humans exclusive of living African great apes. A finding hotly and widely disputed in the West though supported in China by some geneticists and some Chinese paleoanthropologists. If orangutans and humans are less derived relative to African apes (look more like the ancestors of great apes and humans than known African apes do), and chimpanzees diverged from humans before orangutans, but speciated from humans after orangutans, both camps are thus reconciled in a parsimonious manner.

CHAPTER 3

Interpreting Fossil History and Genetics of the Genus Homo

(Image credit La Opcion de Chihuahua) Human skull Mexico 12,000 years old

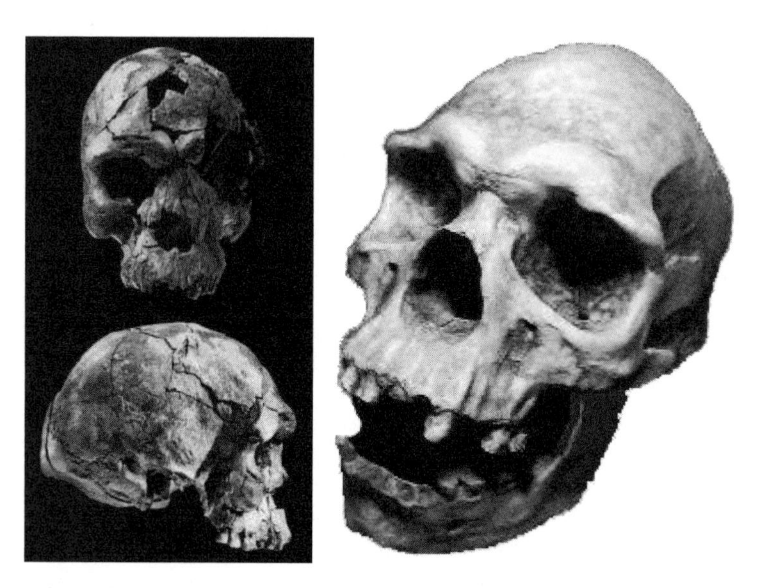

Photos left top and bottom Herto = Homo sapiens idaltu. Right Pintubi 1 1800's Australia.

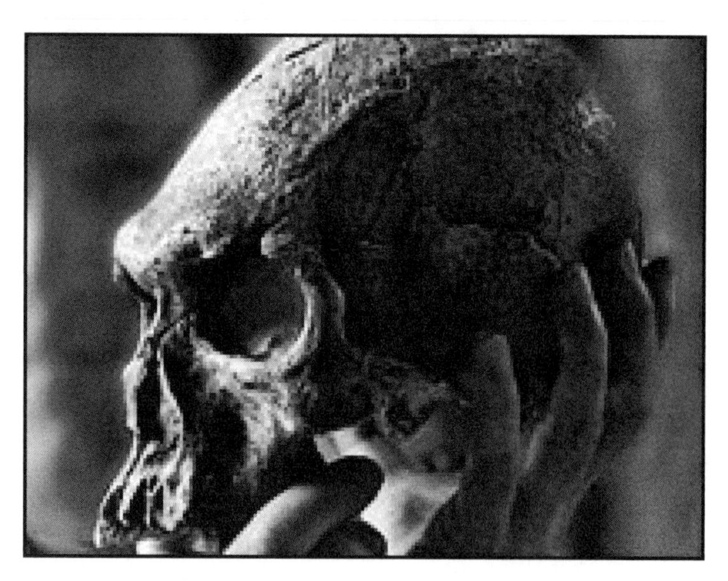

Photo Credit BBC News. Photo typical Original Tasmanian man looted skull lived in the 1800's Tasmania (Sahul)

Discussion of photos. Three of the four above skulls are widely agreed upon to be fully modern humans (Homo sapiens sapiens). Chimalhuacan man for no other reason than being from the Americas, and the Tasmanian and Pintubi skulls for no other reason than being from Sahul and less than 10,000 years old. Homo sapiens idaltu is considered by some authors to be morphologically close enough to H. sapiens sapiens to not merit a different subspecies. The Tasmanian skull is typical for Original Tasmanian males which are gracile and have a fossil record in Tasmania back into the Pleistocene. Some Original Tasmanian men were of robust morphology type not known in the fossil record similar to Southern Original Australian maritime people. Chimalhuacan man grades gently into other early skulls of Mexico as well as some extant from North America and is not indicated to have been genetically isolated from older very modern appearing skulls from Mexico. As skull at 10,000 years is known from Mexico with more modern traits than seen in Chimalhuacan man and still very similar to the Chimalhuacan skull. Pintubi 1 is a typical man for the Pintubi pre-contact hunter-gatherer tribe of Australia, which today has mixed with other morphotypes. These four skulls form a clade based upon considerable character states (cranial slope eye ridge morphology etc.) which are common in the Americas and Sahul and unique from any living African population. Strongly suggesting that living African morphology is not a likely model for the morphology of any putative population of Africans as founders of all Homo sapiens sapiens outside of Africa.

The last 20 years, it can seem as if the more hominin fossils are found, the less we know about hominins. If so then the genus Homo should be the most difficult and poorly understood of hominins, other than fully modern humans. The origins of fully modern humans, Homo sapiens sapiens or anatomically modern Homo sapiens (AMHS), remains controversial. Wolpoff is in many respects correct to have written than AMHS only exists in the imagination. There is near universal agreement today that living humans are all the same species and subspecies, Homo sapiens sapiens. What biological races people today fall into, or if all people today are only one race as Loring Brace and others have proposed, remains disputed. With David Reich sometimes under criticism for claims that there are different races of humans today, and that these races are different from each other. Versus the opinion that variations within races are much greater than between races, so differences between races are trivial. Sahulians whole genomically are closer to all other Sahulians than any other extant population. Native Americans where known for genomes, are closer to all other Native Americans than to any other group. No lines can be drawn between different Eurasian / African populations which clinally grade into each other in small to undetectable clinal steps. Arguably there are three biological races of humans today, Native Americans, Sahulians, and everyone else as a single race. The three groups also arguably form three morphological clades along the same lines.

In Western human genetics and paleoanthropology it is commonly thought low genetic diversity suggests a population bottle neck (founding by a numerically small group of people). This is one possibility, purative natural selection is another more common possibility except for species nearing extinction such as cheetahs (the model often used for supposed population bottle necks in humans). It was thought that agricultural societies over time become more genetically diverse than hunter-gatherer societies, due to larger population densities and higher risk from diseases. In every tested case, the opposite is true, with ancient agricultural societies having higher genetic diversity than their descendents. Hunter gatherers are diverse in their habits and exposed to diverse parasites and diseases, so are selected for relatively high genetic diversity. Highly specialized populations, such as neandertals and agriculturalists, are under strong purative natural selection so develop low genetic diversity. Because natural selection in such populations weeds out all but the most optimal genes for such highly specialized existences.

AMHS has been defined by traits to distinguish us from archaic humans, Homo erectus, and other hominins. It has been claimed that the ossified chin is one mark of fully modern humanness and first seen in Africa about 120,000 years ago. In some pygmy populations, most people have no ossified chins. Some neandertals had ossified chins. Now we can see the Hualongdong (Xiu-Jie Wu et al 2019) skull from China has an ossified chin, and is about 300,000 years old, of similar age to Jebel Irhoud for which ossified chins are not known. In this authors opinion, the Hualongdong skull is as modern as Jebel Irhoud, and was part of a population that ranged West to East from North Africa to China, Hualongdong being an Eastern grade of the same population of which Jebel Iroud was a Western member of. If correct this explains a lot about the commonality of all people today, with this very wide-spread archaic Homo sapiens population being a primary ancestor of all people today though closer to neandertals than to AMHS.

About 70,000 years ago, at some unknown geographical location(s), the human Broca area greatly enlarged with a corresponding internal impression in the cranium not seen before. This is the only trait which appears to unify all historical human populations and distinguish them from all known widely accepted to be archaic humans. Original Tasmanians, and some historical tribes of Original Australians, have front to back craniums much like Herto Man. Neandertals gently grade into AMHS (Loring Brace), and some individual people today are morphologically mostly within accepted neandertal morphological ranges. A few people today would be thought to be neandertals if found in the right places and time. The occipital bun, a projecting bone at the back of the skull for strong muscle attachments, has been proposed as one diagnostic trait of neandertals. The author has an occipital bun as large as the average for neandertals, though occipital buns are rare in Europeans. Some human populations today have on average larger eye ridges than Homo floresiensis.

The enlarged Broca area may have spread by gene flow quickly through all populations of humans by about 70,000 years ago because it imparted universal advantages (was useful to have in all populations of humans). For example in trade and kinship relationships, to help people survive locally difficult times by being able to join other groups at distance where times were less difficult (due to the increased ability to learn new languages especially in children and teenagers), and being more desirable as mates. The absence of the third molars (wisdom teeth) has been suggested as a trait only seen in fully modern humans. The trait is now known in Asiatic archaic humans such as Homo tsaichangensis, and not outside of Asia in archaics and is more prevalent today in Asians than Europeans and Africans.

In terrestrial vertebrates in general, morphological traits tend to move from East to West and West to East more than North to South and South to North. Because morphological traits are to a large degree adapted to local climates, such as temperature zones. Moving up in elevation is like moving towards the poles in morphological traits. Morphotypes often migrate with their climate zones in cooling and warming trends. Deadly microbial diseases and parasites are more likely to originate in tropical regions than temperate or cold regions. Genes related to immunity to deadly diseases such as rhinovirus, tend to move from tropical areas to non-tropical areas, usually leaving morphology genes behind which are usually not favored by natural selection outside of tropical regions.

The unamended ROoA hypothesis, in which all Eurasians were fully replaced by a small population of migrating Africans within the last 70,000 years, supposed all differences between different non-African populations and living Africans, were due to genetic drift mutations and adaptations to different regions taking place within the last 70,000 years. Most morphological differences between non-African populations and African populations of today come from differences in ancestry, especially differences in Eurasian ancestry reaching back to at least Asiatic Homo erectus. For example, strongly dished incisors, which is associated with cold adaptations such as high lactose and fat in mother's milk, is first known in Homo erectus and spread in Crapina eastern neandertals and some AMHS by founder effects and introgressions. Just as Chinese paleoanthropologists proposed by the 1970's. Natural selection and adaptation are in process at all times and places. Not frozen in time any place such as recent sub-Sahara. Ancient Africans at older than 10,000 years before any genomes are known from Africa, were not the same genetically or morphologically as Africans today. Ngandong Homo "erectus" of Java Asia was found to have continuity with WLH-50 a fully modern human from Australia (Hawks et al 2000) to the exclusion of known African fossil humans.

Gene flow is nearly always in every direction at the same time, where other humans exist to accept gene flow. When there is net gene flow out of some area into another, genes are also flowing back though at a lower rate. There can be net gene flow more one way than another, and negative natural selection against parts of participating genomes. Continental interchanges are generally in both directions, though if a species is not on both continents, then for that species the gene flow will be in one direction at first contact only. Such as when humans first reach places that were uninhabited or rare cases of local human extinction such as below ice sheets or oceans and the region is later repopulated when exposed again. AMHS did not originate any one place. AMHS is a result of increased gene flow rates in all human populations, upon which natural selection acted upon. With archaic ancestral morphology detectable in each region in pre-industrial age populations in those places. It was predicted until recently by most Western paleoanthropologists, even human geneticists after neandertal genes were published by human geneticists (which error David Reich corrected), that all people alive today have no neandertal genes, or genes of archaic Eurasians of any type. In every case where archaic genomes are now known, neandertals and denisovans, they have been proven wrong to their own satisfaction by their own work.

A common source of error in human genetics studies is the notion that whole genomic distances accurately reflect phylogeny. This is true for the last few thousand years, but often is not true moving back in time. Unless we redefine being closely related being suggested by sharing common ancestors with similar traits to two groups, to relatedness being based upon who you have kinship relationships with, but often look nothing like. David Reich in his 2018 book wrote that Congolese pygmies are arguably as genetically diverse as the Khoe-San. He also writes it is likely there was a strong back flow of Asiatic genes into all of Africa at some time, though he does not write why.

Unamended ROoA allowed for only one founding African population as being 100% ancestral to all non-African people of today. And that group was not pygmies. African pygmies and Asian and Sahulian pygmies ("negritos"), had to be completely unrelated with "negritos" obtaining their pygmy morphology independently of African pygmies. It was always thought before they were closely related, as they share very similar morphology, African appearances, and similar religion and material cultures to the exclusion of other types of humans. ROoA proved their point by showing that by whole genomic distances, Negritos were Asiatics, and African pygmies African, or did they?

Recently it has been found that African pygmies and negritos share complex heart related genes to the exclusion of other people. And that negritos of the Andaman Islands are whole genomic South Asians, with no detected Denisovan genes, while negritos in and near Sahul are whole genomically in the same group as large people in their area, including high detected levels of Denisovan genes! There is no solution for this problem proposed yet from within ROoA. A solution is that there was a migration of pygmies out of Africa into Asia and Sahul by sea, around 40,000 years ago. And these migrants became fragmented before the Denisovan genes were introduced into the more Eastern populations. Mixing with archaic pygmies still in Asia who were closely related to fully modern pygmies, until Asiatic archaic pygmies were converted to African pygmies in many traits such as the Homo erectus lower bodies.

Extinct pygmies on Madagascar share more traits with Homo floresiensis than any living people do, as do extinct pygmies on Palau Island Micronesia that lived until about two thousand years ago. The fossil record for archaic pygmies, which extends back 700,000 years on Flores Island for Homo floresiensis, suggests living pygmies and negritos obtained their small stature and other morphological genes related to adaptive advantages to living in dense undergrowth, from archaic pygmies in Asia. With fully modern African pygmy genes then flowing back out of Africa to Asia mixing with peoples they encountered along the way. These Asiatic archaic pygmy genes still being incorrectly classified as African in origin with no reference genomes to prove or disprove their origins. Archaic pygmies when known for leg morphology, do not have Homo erectus grade lower bodies. All human populations alive today have lower body morphology of Homo erectus grade. This is just one example of the fossil record strongly suggesting the morphological differences between different populations today have very deep in time origins, well before African people are supposed to have virtually replaced nearly all non-African people.

CHAPTER 4

Review of the Fossil Record for the Genus Homo

	sapiens		
neandertalensis	helmei	tsaichangensis	
antecessor	heidelbergensis		palaeojavanicus
ergaster	erectus	georgicus	
rhodiensis	rudolfensis		
			luzonensis
		floresiensis	
naledi	gautengensis	habilis	

Chart 1– Named species in the genus Homo with sufficiently known fossil material to reasonably name a morphotype from a holotype in the genus Homo.

Discussion of Chart – The chart roughly moves from older to younger for estimated time of origin from the bottom up and West to East from left to right. Homo georgicus is here a synonym of Dmansi Homo erectus. Distances between "species" on the chart roughly correlate to distances of relationships for oldest known fossils before later introgressions. Blank parts are suggestive of a much larger number of morphotypes missing from the fossil record. As Homo is better evidenced as a biological species than a genus (genus a collection of species more closely related to each other than any other species), dichotomizing charts and charts with lots of arrows give false impressions, especially cladograms. Cladograms are good for grants and press they do not give the right impression for mosaic evolution within a single species undergoing adaptive radiation. Meganthropus has been resurrected as a genus, and previously has been dropped and assigned to Homo, so the single species named in Meganthropus, palaeojavacanicus can again be considered a potentially valid species within the genus Homo. The only "species" on the chart the author considers likely to be polyphyletic (to contain more than one group (clade)), is Homo habilis, especially for the oldest known material. The fossil record for Homo is not complete enough to attempt to make meaningful charts from which emerge clusters indicating clades based upon morphological traits. Reality is every Homo that ever lived placed on maps with a third dimension of time.

The fossil record for the genus Homo is very complex. In part because Homo has by far the largest known geographical range of any hominin genus and there are many more fossils for the genus Homo (not including AMHS) than for all other hominin genera combined. Nor is it always clear if fossils, even skulls, are members of the genus Homo, "Meganthropus", or Australopithecus. Homo habilis is a potentially polyphyletic taxa, which when skulls are known, had close affinities with Homo erectus. It is widely thought, that Homo erectus was more adapted to living in open terrains, and to long distance walking, than prior Homo. Archaic pygmies continue to severely challenge ideas prevailing among most professionals.

In 2016 fragmentary remains of Homo floresiensis were published from Mata Menge on Flores Island (van den Bergh et al 2016), from a Xeric (desert) environment, about 20 percent smaller than the much more recent Homo floresiensis remains. It was predicted by this author on face book groups older Homo floresiensis remains would be smaller and less Homo erectus like, which opinion by the author was strongly disputed on face book groups. Some first professional opinions of Homo floresiensis were that they were "primitive" habilines not dwarfed Homo erectus. At the time this author made such predictions, the idea Homo floresiensis was anything except shrunken Homo erectus had been widely rejected, with much evidence and many papers presented to that effect.

In 2017, Debbie Argue et al published a paper in which it was decided by cladistic analysis that Homo floresiensis was as primitive, if not more so, than any known Homo habilis and not a diminutive Homo erectus. Silencing even wilder speculations than before 2016 such as that Homo erectus had reverse Island gigantism more than once. Javanese Homo erectus introgressive traits in Homo floresiensis have been discussed by this author on face book. It has been noted in the literature, that both Homo floresiensis and Homo naledi, have large numbers of modern to fully modern human traits. The Island gigantism / dwarfism hypothesis predicts animals of the size of rabbits or larger, especially carnivores and omnivores like hominins, tend to get larger on Islands compared to their mainland relatives (Lam (Peter) Bao 2010). The rule was reversed in a failed attempt to make Homo floresiensis better fit into prevailing Western hypothesis.

ARHGAP11B is a human specific gene and the only known human specific gene that promotes basal progenitor amplification and neocortex expansion (Florio et al 2015). Ancestral ARHDAP11B promoted smarter higher energy use (more expensive and effective) brains, but not larger brains by limiting anti hyper-encephaly genes, diverging from extant ARHGAP11B about three million years ago (UCSD human genetics lectures). This suggests the ape sized brain Homo clade including Homo naledi and Homo floresiensis, initially diverged from the Homo erectus clade about three million years ago at some unknown location. Homo floresiensis had fire use and late Homo erectus grade tool use, including composite spears. Homo naledi has no tool association but may not have carried stone tools into the caves where they are found for easier passage. Homo floresiensis in one stroke defeated the widely held belief that larger brain size was driven by tool use in Homo.

The oldest known fully modern human dentitions are about 120,000 years old from Africa and East Asia, probably older in East Asia than Africa. The oldest reputed modern Homo sapiens, is Jebel Irhoud, which had been previously published as an African neandertal. Jebel Irhoud is not agreed upon by most professionals to be a modern human and has much closer morphological affinities to neandertals than to fully modern humans. Jebel Irhoud shows "some" modern human traits not seen in other archaic humans, as do all archaic humans.

Herto man (Homo sapiens idaltu), is the oldest widely agreed upon modern human, from Ethiopia at about 160,000 years old. Older fragmentary remains in the general region show affinities to Herto back about 200,000 years. Herto like other known early modern humans was quite robust with larger more ossified (thick) skulls compared to human populations historically outside of Australia. Some Original Australian tribes historically, and Original Tasmanians, have very Herto like cranial morphology. Paleoanthropologists today do not have literature to distinguish different fully modern skulls reliably by continental origins, forensic anthropologists do with about 90 percent reliability. When Herto is examined by forensic anthropologists, Herto clusters with Sahulians not Africans (Tim White lecture 2017). Herto is another example of morphological differences between populations today being deeply seated in time. In Southern Africa, the fossil record for this time frame is not as good as in Eastern Africa. About 70 thousand years ago, Khoe-san like morphology replaced more Herto like morphology over a short period in Southern Africa. Khoe-San like dental morphology is known farther North in Africa back to about 100 thousand years.

Southern Africa was well connected by sea ice to Tasmania 140,000 years ago. In this period the sea ice was along with beaches were the most favorable places to live on Earth, for people able to make small boats and clothing, which is indicated to be the case at that time. The largest masses of mammal flesh in the World were on the Antarctic sea ice in this time period, seals, along with diverse other food resources unlike today. Antarctic sea ice today is not an accurate model for the sea ice of greater Antarctica, which was much warmer than now where the ice met the open sea. It is unlikely people could have killed whales that long ago, though they would have scavenged dead whales. People living on the sea ice could have lost the ability to make stone tools, as with some AMHS living in heavily forested areas, using instead bone hide and sinew, and later wood tools.

Here Homo ergaster, or "African Homo erectus", is considered a synonym of Homo erectus following Tim White and other authors. There is a common trend of Homo populations having smaller brains moving towards and in East Asia than their more Western grades. This has led to the incorrect conclusion Homo heidelbergensis did not exist in Asia. The Narmada cranium from India, in this authors opinion, has closer affinities with European Homo heidelbergensis and has been published as Homo heidelbergensis by some authors. Other Asiatic fossils are also very similar to Homo heidelbergensis. Brain size is more likely a function of environmental factors than a measure of linear evolution of hominins, or Homo, from "lower" to "higher" forms. Small brain size relative to African and European relatives such as neandertals, has also led to some archaic Homo sapiens in Asia being incorrectly assigned to Homo erectus.

Homo erectus was very successful and ranged widely from Africa through Eurasia. Homo erectus at Dmansi is the oldest known snow bound winter adapted hominin and the oldest known Homo from Eurasia at about 1.8 million years. In China stone tools are known at 2.1mya, and from the Siwalik of Pakistan tool made cut marks are known on bones at about 2.1mya. The high morphological diversity of the five Dmansi skulls have reinforced a common perception that far too many species of hominins have been named. Homo erectus is known from Africa from about 2.1 million years and is probably of African origins. Dmansi Homo erectus have smaller brains than any other known fossils widely agreed upon to be Homo erectus, and in general have more "primitive" traits than known African Homo erectus. It may be that Homo erectus began in Eurasia or began both in Eurasia and Africa at the same time in the same population which ranged from Africa into Asia. Homo erectus is not generally recognized in Europe, a jaw with Homo erectus morphology has been published from Serbia. When Homo erectus morphology went extinct is not well agreed upon, with estimates lately running from 100 to

30 thousand years ago. In China there are teeth known from around 20 thousand years ago intermediate morphologically between fully modern humans and Homo erectus.

Homo from Africa which do not fall within Homo sapiens or Homo habilis, are all likely to be grades of Homo erectus or too fragmentary to reliable assign to new species. Tim White argues that Homo naledi (Berger et al 2015), an ape sized brain Homo found in two cave locations in South Africa dating about 250 thousand years old, are a grade of Homo erectus. Other Southern African fossil hominins are from wetter periods than Homo naledi, Homo naledi could be more of a desert adapted hominin with a very long history than associated with the late date where found. Homo naledi may have left their stone tools outside of the caves, to make the difficult climbs into the caves easier and / or to free their hands to carry supplies of perishable materials such as plants and hide bags. Berger et al have proposed intentional burials for Homo naledi, there is little support from professionals outside of South Africa for this idea. Homo naledi hands are quite modern, hand bones are poorly known in Homo erectus. Suggesting fully modern humans get both their legs and their hands from Homo erectus and Homo erectus closely related Homo, and not from archaic pygmies.

Homo tsaichangensis (aka Penghu) to the exclusion of all other known hominins, share many dental character states with Javanese Gigantopithecus (Marc McMenamin 2015). Homo tsaichengensis is known only from a lower jaw dredged from the ocean near Taiwan in association with Pleistocene large mammals. Penghu has obvious close affinities with the jaw from a cave in Tibet identified by proteomics to be Denisovan. Penghu is missing the m3 (wisdom tooth) from birth, and the teeth are heavily worn, possibly in part from eating sandy clams. Denisovan teeth from Denisova Cave are in the adult teeth all upper teeth, so do not compare well to the lower teeth of the Tibetan and Taiwanese denisovans. Denisova Cave teeth are like the lower Denisovan jaws quite unique among hominins, and suggestive of a more foliverous (leafy) and insectivorous diet compared to other Homo except one lower jaw in early wear state of Homo habilis from Africa OH 7 (Leakey et al 1964). Louis Leakey published OH 7 as a direct ancestor of modern humans, naming a new species Homo habilis, with Homo erectus as a dead-end side branch. Denisovans are not a taxa and were not meant to be by the geneticists that named the group, denisovans are a cluster of genes not known from outside of Asia which are present in longer unbroken sequences than known today for neandertals in living humans in Oceania and Sahul. This can be because these genes came into living humans more recently than neandertal genes have, or because these genes are selected for together as groups by natural selection as they need to all be present to be advantageous. When archaic humans have founded modern human populations, such as Loring Brace found to be the case for neandertals founding some Native North American groups , the Jomon, and archaic Egyptians founding modern Egyptians, archaic gene sections will remain longer for a longer period of time than if they came in by introgressions. This is because when the population was mostly archaic and already mixed, archaic genes were more likely to be coupled with other archaic genes in mitosis. Cutting the suggested timing of "introgression" by gene sequence length nearly in half. C. Loring Brace first proposed that Neanderthals were ancestral to early modern humans (EEMH or early European Modern humans then known as Cro-Magnons), in his 1962 paper titled Refocusing on the Neanderthal Problem" in the American Anthropologist journal.

Neandertals were the first archaic humans to be known and published from fossils. Linnaeus 1758 first named the genus Homo with one species sapiens (Man wise). The genus Homo and species sapiens have precedence over any other genus or species found to be a synonym of them. It might be more correct for example to write Homo sapiens race heidelbergensis variety neandertalis. Neandertals were probably founded by Homo heidelbergensis from Africa, or African origin Homo heidelbergensis in Europe, or both, and thus have been renamed Homo heidelbergensis neandertalis by one Spanish author. The oldest known fossil genomes of hominins are from Homo heidelbergensis like neandertals sharing mt DNA with denisovans in Spain at about 430,000 years old (Meyer et al 2016). Some neandertals have ossifications on their chins. A few early neandertal females are more modern human in appearance than any other known fossils their age or older. Neandertals are relatively well known as fossils because they often buried their dead, often lived in caves and because Europe is better explored than any other place for the neandertal period. Neandertals were robust cold adapted humans with heavy muscles and strong sprinters. It is widely thought neandertals killed only from up close. This is unlikely, as very advanced javelins (throwing spears) are known from a peat bog in Germany, dated to neandertal times at about 300,000 years ago (Milks et al 2019). It seems unlikely neandertals would have devolved losing their ability to make or use javelins. Some neandertals used marine resources, ate rabbits, or are indicated to have been vegetarian based upon dental calculus (plaque) studies, while others appear to have been functional carnivores. One study suggests neandertals ate less meat than has been thought based upon biochemical studies, because partly rotted meat leaves a stronger chemical signature for meat in the diet than fresh meat.

Neandertals maintained their morphology through natural selection, not genetic isolation. Wolves in this part of North America showed considerable morphological introgression from domestic dogs in the authors youth, and now through purative natural selection are not distinguishable morphologically from wolves of 200 years ago. Morphologically pure wolves coming down from Canada, and interbreeding with wolf dogs, to yield morphological wolves within a few generations. Once more modern tools reduced the need for neandertal morphology enough the cost of such morphology was no longer justified, it became rare in favor of introgressed modern and fully modern Homo morphology (Loring Brace). Neandertals are not known over six feet tall, though most neandertal men weighed hundreds of pounds lean. Later neandertals like the modern humans they mixed with, on average had larger brain sizes than populations today. In SW Asia, neandertals gave way to modern humans and then the modern humans gave way to neandertals in some places following their climate zones.

Ideas about neandertal inherent inferiority rely upon the false concept of linear evolution, and have a long history going back to remarkably ignorant interpretations of neandertals by 19th century scientists. Often not crediting neandertals with the power of speech or upright walking. Club wielding nearly mindless brutes. It now appears late neandertal art has often been misdated due to assumptions neandertals did not make realistic art, to after neandertal times. And neandertals became artistic as they began to more freely mix with other types of humans with increased trade levels around 48,000 years ago. As well as adopting many modern human tools. Neandertals had a very large range through most of habitable Northern Eurasia but were low population density so existed in relatively small numbers compared to people in parts of Africa and more Southern Asia. Species go extinct, not populations or races within species like neandertals. Neandertals were the longest lasting non-Homo erectus morphotype known so far, and they live on in all people today genetically and many Eurasians and Native Americans morphologically.

Archaic pygmy fossils include Red Deer Cave people, from about 14,000 years ago in SW China (Xueping et al 20112). Red Deer Cave people had cranial morphology in some ways similar to a Dmansi skull, and similar to some living Malaysians. Red Deer people legs were not at all like other modern humans or Homo erectus, and well adapted to tree climbing as with other archaic pygmies. Curnoe et al 2015 published a quite archaic partial femur a little younger than Red Deer Cave people from Yunnan China showing mixing with fully modern humans. Homo floresiensis. Homo luzonesis from the Philippines at about 40,000 years of age, with close affinities to Homo floresiensis. Palau Island Micronesia "pygmies" with close affinities to Homo floresiensis that lived until about 2000 years ago (Lee Berger et al 2008). Some fragmentary remains from the Narmada Valley in India which are quite different than fully modern humans. Extinct pygmies from Madagascar with more affinities with Homo floresiensis than any historical humans. Suggesting a very wide range over a very long period with high morphological diversity between different grades. Living pygmies are unique from other living humans morphologically in many ways shared with archaic pygmies. Pygmies often do not have ossified chins, archaic pygmies are not known to have had ossified chins. Pygmies and archaic pygmies are of shorter stature relative to Homo erectus and above grades. Both have strong arboreal (dense forest) tendencies and are well adapted morphologically to moving through dense vegetation. Both are neotonous relative to their Homo contemporaries (look more like children than other people, large appearing high set eyes, low sexual dimorphism etc.). A hominin rhinoceros butchering site in the Philippines is dated at 709,000 years ago (Ingicco et al 2018) from unknown humans. The oldest known humans of the Philippines are archaic pygmies Homo luzonensis.

Fitzpatrick et al 2008 have published a paper refuting the prior claims of Lee Berger et al 2008 that Palau fragments represent a dwarf population. And conclude Palau fragments represent small individuals of normal sized Palau populations. They do not address the remarkable morphological similarity of the fragments to Homo floresiensis, or that these diminutive cluster of individuals could have arrived to Palau already as archaic pygmies as exotic immigrants coming before or after agricultural Palauans and introgressing with them to result in some Palauans having similar Homo floresiensis like morphology. Berger et al 2008 assume diminutive Island humans with Homo floresiensis like morphology to be a result of reverse Island Gigantism, in vogue then, and not widely accepted to be the case for Homo floresiensis since 2017. In the skull shown by Fitzpatrick et al 2008 of normal size a very unusual not at all fully modern human appearing cranial slope and other archaic appearing morphology is present. Identifications which ignore morphology in favor of what fossils should be based upon dates is now dangerous in Asia and Oceania for humans. As are assumptions Philippine residents of 708 thousand years ago were "dead end lineages".

Homo heidelbergensis likely originated in Africa and are the primary ancestors of neandertals and a main ancestor of Homo sapiens. They were the first humans to reach fully modern average brain size. It appears that Homo sapiens also had some Asiatic Homo erectus ancestry which never passed through any Homo heidelbergensis grade. Homo heidelbergensis are arguably in some cases Homo erectus, in others archaic Homo sapiens, and in others early neandertals, and may not be a valid taxa. Instead a chronospecies / transitional grades moving from Homo erectus to Homo sapiens. Homo heidelbergensis was first named for lower jaws found in Germany in 1907 and was one of the early known non fully modern humans. Homo heidelbergensis is recognized from about 700 to 300 thousand years ago. Identifying taxas and progressions solely based upon brain size, and cranial bones related to brain size, is unlikely to create accurate phylogenetic trees.

One hypothesis of Homo heidelbergensis is they evolved from Homo antecessor (Eudald Carbonell 1997) with a cranial capacity of about 1000 to 1150 cc of Europe at 1.8 to 1.2 million years. Like with the Taung child, some conclusions made about Homo antecessor being directly ancestral to fully modern humans were based upon work with juvenile material. Homo antecessor is likely to be an invalid taxa and synonym of Homo erectus with cranial capacities of 800 to 1100 cc or Homo heidelbergensis with cranial capacities of about 1250 which is about 90 percent of that of fully modern humans. Homo antecessor had occipital buns as did neandertals and some living humans. An analysis of the Gran Dolina- TD6 previously unpublished Homo antecessor teeth (Maria Martinon-Torres et al 2019) finds Homo antecessor dental morphology suggests H. antecessor was basal to Homo sapiens neandertals and denisovans. That Homo antecessor is distinguishable from Homo erectus, and shared many traits with European hominins, does not suggest H. antecessor could not be a European grade of Homo erectus closer to the ancestry of European hominins than Asiatic and African Homo erectus through continuity in Europe. It is likely first Homo heidelbergensis entered Europe from Africa mixing with Homo antecessor in Europe and then neandertals migrated to Europe from Africa mixing with hominins already present in Europe. The known age of Homo antecessor supports assignment of the fossils to the genus Homo as do some mosaic morphological traits.

Yuan et al 2017 "Using improved methods and public data, we have revisited human evolution and derived an age of 1.91-1.96 million years for the first split in modern human autosomes." These authors have found in the genes of people today, a divergence that took place at nearly two million years ago and did not take place in Africa. The oldest known human fossils from Eurasia (in Asia close to Europe) Dmansi Georgia 1.8 million years. The oldest tool use evidence and tools from Asia, Pakistan and China 2.1 million years ago. It was not just Homo erectus, which adapted to extreme cold climates around 600,000 years ago, contributing cold adapted genes to neandertals and denisovans. Early Homo erectus storming into Eurasia around 2 million years ago split the ancestral family tree of people today into an African and Eurasian branches. With neandertals and Homo heidelbergensis coming up into Eurasia from Africa much later, to mix over time with Eurasians including archaic pygmies to yield fully modern humans. The author does not agree with Yuan et al that African features found their way to negritos via neandertals and denisovans. Genetics without paleoanthropology does not provide information about the origins of genes. Archaic pygmies exchanged genes with neandertals, giving their negrito descendants commonality with neandertals. With fully modern pygmies coming out of Africa founded by Asiatic archaic pygmies, to mix with archaic pygmies in Asia giving negritos their African appearance directly from Africa in recent times (around 40kya). Into Africa and then back out of Africa.

The oldest hominin genetic material known is from proto-neandertals from the Sima de los juesos in Spain dated at 430,000 years. Showing clear affinities with Homo heidelbergensis justifying the taxanomic reassignment of neandertals to Homo heidelbergensis neandertalis. These early neandertals had mt DNA closer to that of denisovans unlike all other neandertals known for mt DNA (Meyer et al 2016). Later neandertals also had very deeply divergent mt DNA (Peyregne et al 2019). Suggesting considerable introgression in neandertals after they migrated out of Africa as archaic Homo sapiens of Homo heidelbergensis grade. So many deeply divergent mt DNA in small neandertal populations, gives credibility to the reported deeply divergent mt DNA in fully modern humans in Australia reported by Adcock et al. All fully modern human mt DNA of great age known is from permafrost areas (with the possible exception reported by Adcock et al from Australia), representing highly specialized AMHS, with other than Sima mt DNA closer to fully modern human mt DNA than to denisovan.

If more mt DNA samples were known from fully modern humans of similar age to the one reported by Adcock et al, this author predicts they will sometimes have deeply divergent mt DNA, both in and out of Africa. A deeply divergent Y chromosome is now reported in living humans in Africa (Haber et al 2019). With so many very ancient ghosts and genes now known from Africa, some likely to be archaic even from taxa such as Homo naledi (the deeply divergent saliva gene for example), it is not easy to explain why these genes did not travel into Eurasia as they originated in populations thought to have been extinct well before 120,000 years ago in some cases. However, if natural selection acted upon genes leaving Africa through ordinary gene flow, instead of populations of migrants essentially replacing archaic Eurasians of all types, then these deeply divergent genes could have more easily been taken out by natural selection being better adapted to Africa than to Eurasia.

CHAPTER 5

Giants and Exobiology

Photos Top Homo erectus palaeojavanicus (Meganthropus) from Java Indonesia Sangiran VIII and an artist's concept of historical Patagonian giants South America. Sangiran VIII was a true giant, the largest known undisputed hominin of any time. It is often claimed certain African Australopithecus had the proportionately largest eye ridges of any hominin. This is incorrect Sangiran 8 and Sangiran 17 had much larger eye ridges absolutely and proportionately compared to any known European or African fossil hominin or ape.

No human populations today average over six feet tall. No neandertals are known over six feet tall, no denisovans are known for height. Ona giants of the Tierra del Fuego of Patagonia (the Southern tip of South America) from 1520 to 1899 have been reported to include individuals well over seven feet tall. Bergman's rule was meant for species within a genus but has been extended to populations within the same species. Suggesting that larger bodied populations are found in colder climates, which can partly explain why the Ona giants were so big. No Ona giant skeletons have been published, but there are photographs and numerous first-hand accounts of them. Ona giants went extinct in about the early 1900's. their neighbors were small people. Ona giants have been estimated to have been about seven feet tall on average. More likely they averaged about six and a half feet tall. Some Native North American hunter gatherers were reported as averaging over six feet tall in early contacts. Ancient accounts of some Northern European tribes suggest they averaged over six feet tall. Inuit legends report the last Greenland Viking populations were a mixture of giants and dwarves, a pathological condition from excessive inbreeding. Excavations of very late period Viking graves now support this. The traveler Dr Frederick Cook visited the Ona in 1897 publishing photos of them. Cook reported the Ona men averaged six feet tall and often were over seven feet tall. It is likely the Ona of the 1500's averaged much taller on average than the Ona of the late 1800's. There are reports in Spanish language, some with photos, that the Ona are not an extinct tribe. It is likely the Ona had been reduced in average height by 1897, with sone still expressing their original size. Through mixing with their small stature neighboring Native American tribes, Europeans and Africans.

Some Meganthropus were very heavy bodied, weighing 800 or more pounds and over seven feet tall. By far the largest known hominins, from the Pleistocene of Java Indonesia. No hominin leg bones are known massive enough to support heights much over seven feet tall in adaptively fit individuals, and no leg bones are known for either Meganthropus or Gigantopithecus. Many fully modern archaic humans cared for people that could not have survived, including pathological giants who could not walk, or could not walk far. The ancient Egyptians used giants and dwarves to work gold, it is thought in part so they would be easy to identify if they ran off with gold and because they could not run so fast or far. Today non pathological humans do not reach much over seven feet in height. The on average tallest populations today are in Africa, such as the Maasai. The Amazon tribe, a branch of the Scythians, were reported to have in ancient times included some large stature women, though they are poorly known archaeologically so their average height is not known. Some types of combat can select for heavy / tall individuals. There are many legends of large bodied individuals from the Dark Age of Europe when heavy armor difficult to penetrate with weapons of the time was used.

The taxonomic position of Gigantopithecus, some of which were by far the largest primates ever known, remains problematical. A past director of the American Museum of Natural History, and great anatomist, believed Gigantopithecus are directly ancestral to modern humans. Weidenreich believed that modern humans are derived from giants. The main reason given for abandoning giant humans as ancestral to modern humans, was on principle not based upon morphological arguments. In that larger taxa usually do not give rise to smaller taxa in mammals. There are of course many exceptions to this "rule", for example gomphotheres (a type of ancient elephant), steadily declined in size for millions of years in the Miocene of North America. The author does not believe that giant humans were primary ancestors of present-day humans. The author however does strongly suspect Gigantopithecus was a stem hominin, not a great ape, and certainly not a stem aberrant pongine (orangutan).

Gigantopithecus was first tentatively moved out of hominins in 1972, in the belief that early hominins should be African not Eurasian (Gigantopithecus is known from Germany and widely in Asia, not in Africa). Oldest Gigantopithecus age uncertain and it is not known how old the oldest remains are. The movement of Gigantopithecus out of the hominins was finalized based upon the shared lingual groove, a valley in the teeth, between Gigantopithecus and orangutans, and has been given little attention since the 1970's as to taxonomic position. In large part because not a lot of new Gigantopithecus fossils have been found. The lingual groove is likely to be a plesiomorphic trait, one shared from common ancestors of Gigantopithecus and orangutans and lost in other great apes / humans. In modern systematics, this is not adequate evidence for such a revision, just as beliefs about where geographically taxa should occur is not valid for assignments in systematics. It is common in mammals, especially arboreal mammals as Gigantopithecus, for there to be isolated appearances of taxa on other continents than they are generally known from. Such as stray occurrences of Asiatic arboreal squirrels in North America, due to interchanges in the Miocene between Asia and North America through forested corridors connecting the continents. Arboreal mammals are rare in North America from the times of great apes in Asia, they could have been present in North America and missed in the fossil record.

There has been speculation, if for no other reason than the largest Gigantopithecus could not likely have walked very far on two legs due to their great body weight (well over 800 pounds), that at least some Gigantopithecus were quadrupeds of some sort. Only jaws and teeth, in large numbers, are known for Gigantopithecus, no postcranial material at all is published or known. Most Gigantopithecus fossils known were about the same size as people today, though they are generally presented as a race of giants in popular literature. If the remarkable level of shared morphology between the morphological Denisovan Homo tsaichangensis and Javanese Gigantopithecus is through close relationship or introgression, how likely is it that Gigantopithecus is a great ape not a basally divergent hominin? Not too likely, by either introgression, which sometimes can take place in monkeys at such distances, or by HGT which would not normally transfer so much morphology. If however Gigantopithecus were basally divergent hominins, then it would be much more likely Gigantopithecus and Homo erectus could interbreed to yield Meganthropus, which in some specimens fall into great ape range morphologically. Or that Gigantopithecus could introgress with Australopithecus to yield Paranthropus, which inexplicably in current models without invocation of unusually high levels of convergence, shares many traits with Meganthropus. The four oldest genera of likely hominins are not more likely hominins than Gigantopithecus, which was excluded in the false belief that very early hominins should be chimp / Australopithecus like and African.

Gigantopithecus lower canines have a honing process which wear them down by occlusion with the upper teeth. This is known only in hominins otherwise, not in great apes. Gigantopithecus ate a lot of bamboo shoots, which explains why their dental morphology is so different from any known primate, they were adapting to a very coarse diet in their dental morphology. If we are to accept no less than four genera now of likely hominins, all with many great ape traits, then it should be no harder, given the evidence, to accept Gigantopithecus as a likely basally divergent hominin. Frayer D. W. 1973 in a time when Gigantopithecus was very actively studied, unlike all later decades, wrote "The total morphological pattern of Gigantopithecus mandibles is more similar to Australopithecus than to P. gorilla." And "G. blacki might be considered an aberrant hominid rather than an aberrant pongid." In 1973 when this was written, the term hominid had the meaning to the term hominin today.

Nor was Homo tsaichangensis, in many ways an intermediate form between Homo erectus and Gigantopithecus, known in 1973, very strong evidence Frayer was quite correct. All this suggests that the Denisovan genetic signature is very deep, coming from a basally divergent hominin in Eurasia, Gigantopithecus. That one dental protein is shared along with the lingual groove, with very deep divergence suggested, between Gigantopithecus and orangutans does not refute this notion. Some Lufengpithecus were closer morphologically to relic Asiatic Homo than any other known great apes. And some Lufengpithecus had dental morphology closer to orangutans than any non-orangutan known. In mosaic evolution there is no issue with orangutans sharing a dental protein with Gigantopithecus, and Gigantopithecus being a basal hominin, as the divergence suggested is well back into the times when great apes and hominins were likely freely interbreeding to yield fertile offspring (one morphologically diverse species). Nor is there any evidence as to when exactly Eurasiatic great "apes" other than extant orangutans stopped having gene flow with hominins from genetics, as no Eurasiatic great ape genomes are known. If similar to African apes in timing of divergence, it would be around 7-9 million years ago. If Gigantopithecus diverged from some stem hominin along with Paranthropus, both would likely be able to yield fertile offspring with people today and likely to have the hominin chromosome count unlike all extant great apes.

Though the idea of aliens in spaceships visiting people is very popular today, there is no morphological or genetic peer reviewed literature on aliens. Panspermia is the hypothesis that life on Earth originated on other planets and drifted to Earth through space. Which potentially includes animals especially tardigrades capable of surviving space travel. Other types of animals or their eggs could potentially survive frozen in ice such as in comets. Because mountain tops experience lower temperatures than basins, the first liquid water, and probably the first life, on Earth was not in basins as widely thought, and was instead on mountain tops. Virus in some cases have no known matches on this Earth for most of their genomes.

Through HGT (horizontal gene transfer between different taxa including sometimes distantly related ones), humans could potentially have many genes in us from other planets, brought to us by microbes even small animals drifting through space. A lot of the human genome is viral, and viral genes sometimes come into play being expressed in our genomes in diverse ways. Which could give some commonality to higher organisms like humans with organisms on other planets. If such genes effecting our morphology are present, they would have come to us long ago to wait to be expressed in our phenotypes when environmental conditions became right. Animals on other planets also could become human like through convergent evolution. Such "humans" would be more distant genetically from us than any animal on Earth, with no chance of hybrids only a remote chance of HGT. "Humans" sharing HGT genes with us also would not possibly yield offspring with humans, fertile or infertile, and would be distant from us genetically. Genes can be arranged in many ways in a genome for the same function, and there are different genetic solutions to similar problems.

In the Americas a unique type of sexually transmitted disease began in dogs. These cancers have spread in dogs around the World and have elements of the first dog's genome the cancers came from in the cancer genome. It is possible alien organisms could use Earth life for food or for a medium to live in or on. There is no evidence of any such organisms on Earth, nor is it clear if virus are Earth life or not. Virus are a driver of adaptation in that they bring genetic solutions from distantly related organisms which the recipient might not have been able to solve on their own. Adaptation to diseases is also assisted by viral HGT origin genes in bringing experience with diseases from other animals that have developed resistance to diseases. If panspermia is correct, then many diseases might exist on different planets and we already have resistance to some of those diseases. If life on other planets does not share genes with us from common sources, then we would lack resistance to alien diseases. Often the alien diseases would not be adapted to infect us. In space travel it is easier to transmit genes than whole organisms.

Tardigrades are tiny animals with eyes mouth parts and legs known today and from the Cambrian. Tardigrades likely could survive frozen in space, certainly frozen in ice in space. If lower order animals experienced a lot of undetected HGT early on in the history of life on Earth, or in space, it could incorrectly create the appearance that animal life on Earth is of a single origin on Earth by a sort of genetic blending process. In a similar manner to people today all being the same subspecies by a blending process of increased gene flow rates between archaic humans and Homo erectus which may not have all been the same biological subspecies. About 1/6 of the Tardigrade genome is detected to be from distantly related organisms through HGT, by far the highest detected rate for any animal (Boothby et al 2015). In the early Earth life was under greater stress by severe environmental strain than since mammals have existed on Earth. The paper suggests animals under stress more actively exchange genes through HGT. A 2016 paper suggested some of the bacterial "HGT" sequences were too long to be HGT and were more likely from contamination of the sample. The splicing of entire or very long bacterial gene sequences into Tardigrade genomes might have taken place under conditions of severe stress. Bacteria routinely swap packages of genes, from small to large, with other bacteria during times of severe stress from environmental change. From a cladistic perspective, humans are a type of bacteria as we began as bacteria. Over time higher animals have found the need to protect their very complex and fragile genomes from frequent genetic exchanges with other species.

CHAPTER 6

Relationships of Humans to other Great Apes and Primates

9 Gorilla	12.5 Pan			5.7 Homo		2 Pongo
9 Chororapithecus			3.5 Kenyanthropus		2.6 Paranthropus	
		5.8 Ardipithecus	4.2 Australopithecus		2 Meganthropus	
7 Sahelanthropus			6 Orrorin		? Gigantopithecus	
			7.2 Graecopithecus			
					10.5 Lufengpithecus	
12 Pierolapithecus		12 Danuvius				
			13 Dryopithecus		12 Sivapithecus	

Chart 2 – The named genera of the family Hominidae. Chorapithecus is considered a synonym of Gorilla by the author. The closer the genera are on the chart the closer the relationship. Generally the author considers the named genera closer to biological species than genera, especially lower on the chart. The top line is those relics of the Hominidae which have happened to survive until today. The number is the estimated years, in millions, in which the genus is first known in the fossil record. Homo is dated to 5.7 million years because the author considers the fossil track ways from Crete to be diagnostic for the genus Homo.

The more distantly related two animals are, the less likely they are to have beneficial gene flow by HGT. Viral diseases common to distantly related animals, such as birds and mammals, can be their own cure through HGT. For HGT to take place, it must be possible for a microorganism to infect both the source of the HGT DNA and the destination. For examples feathers might be an advantage to some mammal, though no mammals are known to have ever had feathers. It may be feathers will not work in a mammal genome without problems preventing actual feathers. Or feather related genes would cause so many negative issues in a mammal genome they would never be advantageous for a mammal. Or it may be that the gene complex to yield feathers is too spread out in the genome for virus to transfer from birds to mammals.

In general paleoanthropologists and paleontologists are not trained in HGT and / or introgression and are weak in the subjects. A couple of years ago a masters dissertation was published in South Africa on introgression for paleoanthropology, this is the only publication in paleoanthropology that the author is aware of giving basic instruction in introgression. Paleoanthropologists are now widely aware of introgression, however, the topic has generally been avoided / ignored until the last few years under pressure from results from human genetic studies. Jacques Ruffie has written useful books with more modern perspectives on human genetics including introgressions for some time, including a book with Laurence Garey in 1983 "The population alternative: a new look at competition and the species". Paleoanthropology elected to stick with 19th century origins typological methods because introgressions and HGT are more difficult to learn and teach and have not thought to be useful until the last few years. In the cladistic approach, everything is arranged into neatly branching lineages whether the data fits such a model or not. The synthetic approach which includes cladistics is more useful for human and primate studies.

There is a detected HGT event into chimpanzees from baboons detected, and not found in other great apes or hominins. The author has attempted to detect baboon morphology in chimpanzees with no success.

Chimpanzees often eat baboons, especially young ones, and sometimes become infected this way by baboon retrovirus, in a similar manner to humans becoming infected with AIDS from chimpanzees. Baboons were probably restricted to Southern African when baboon genes entered the chimpanzee gene pool by HGT, which places chimpanzees into Southern Africa at the time of the HGT event.

Known hominins when they move quadrupedally, except some knuckle walking Australopithecus, are palmigrade (they walk on their palms). No other known great apes or Hylobatidae (gibbon family) are palmigrade on the ground. Terrestrial monkeys, such as baboons, and some early apes not closely related to the hominids (great apes including humans), do / did walk on the ground quadrupedally using their palms. This could be a case of convergent evolution, or of hominins reverting to an older form of locomotion in their history. Or it could be a case of morphology transference by introgression or HGT.

Moving back in time ancestral populations become closer morphologically and genetically to related groups. While groups may be different biological species today, if you move back in time far enough, they were the same biological species. Within biological species there are different populations with different morphologies, though often there will be some overlap and normally each morphotype will grade into other morphotypes in gentle clinal steps if the populations are members of the same biological species. Once populations speciate, usually they will diverge morphologically rather quickly in terms of geological time, and there will be a morphological gap between the two species, which is not filled by intermediate forms. An exception which is rather common in primates, is hybrid zones. Hybrid zones can result from two different biological species forming populations of hybrids, through inter-specific introgression. Or hybrid zones can be relics from before the populations speciated. Early likely hominins, which show mosaic mixtures of ape and hominin traits, can be either of these types of groups. Results of mating between apes and hominins, or relics from before apes and hominins speciated, or resulting from introgressions of non-great ape primates.

Populations being lines moving back in time to eventually split from other neat lines, is incorrect and is based upon the false concept of linear evolution. While the species concept is quite valid and well defined in biology, though not at all defined or valid in paleoanthropology which has been quite independent of other biological sciences. Moving back in time it is more useful to think in terms of fluids splitting, sharing connections, spraying droplets into other pools etcetera. Gene flow is not like a rigid solid which once separated can't rejoin. It is fluid in movements with increased genetic isolation over time and full genetic isolation seldom a reality.

An introgression into bonobos from an unknown great ape is detected, exclusive of other chimpanzees (Pan troglodytes) and other living great apes and hominins, at about two million years ago (Kuhlwilm et al 2019). The unknown ape is extinct, and there is no evidence of the ape's morphology from the detected genes in bonobos. It is likely that the more human like morphology and culture of bonobos, relative to Pan troglodytes, is from this introgression. It has been predicted being more bipedal and otherwise more human like should impart increased ability to make and use tools. Among chimpanzees, Pan troglodytes are the great tool makers and users, and fighters, bonobos are lovers who rarely make or use tools. Considerable evidence of introgressions of African apes into Western hominins (especially Australopithecus), and of Asiatic apes or stem hominins into Eastern fossil Homo, are likely to have been introgressions from great apes into hominins. Following the model of the detected great ape introgression into ancestral bonobos, just before the Congo River became large partly genetically bonobos from Pan troglodytes.

With almost no fossil record for great apes in Africa, and no fossil record for the Hylobatidae outside of Asia (gibbons and relatives or lessor apes, which are a relic of the groups that led to great apes and humans), not much is known about the relationships of African apes to humans. African apes could be of African / European origins, as suggested by Peirolapithecus from France for gorillinae, or African apes may all be of Eurasian origins in the Miocene at some time. The Pan tooth from Kenya at 12.5 million years suggests African apes were in Africa at that time, though they might also have been in Eurasia at the same time. Lufengpithecus subnasal morphology is shared with African apes but not with other Euraisian apes, suggesting Lufengpithecus is close to the ancestry of the more derived extant African great apes. In the Miocene apes in general may have gone extinct in Africa, under severe pressure from both deforestation and competition with monkeys. Apes developed tolerance to alcohol, monkeys have not, allowing apes to eat fallen and hanging rotting fruit inedible to monkeys.

The fossil record of interchanges between Asia and North America for arboreal mammals, along with the fossil record of arboreal mammals in Asia, suggests that Asia was the primary refuge of arboreal mammals in the late Miocene. Apes might have been present in Africa, in the late Oligocene up until today, possibly even great apes from the Miocene. And some very large gaps in the fossil history of Africa have been filled in recent years. Evidence is good at some span of time from 8 to 14 million years ago no primary hominin ancestry was in Africa. Nor are ape like African monkeys such as Proconsul widely accepted as ancestral hominins today unlike in the past when all things African were pounded into a line leading up to humans.

Pliopithecoids were a group of sometimes bipedal stem monkeys that sometimes became quite ape like, some of them were ape size though most were small. They had a wide range with high diversity from Africa into Europe and Asia though they are rare as fossils. It has been proposed Pliopithecoidea are ancestral to great apes. This is unlikely, in that pliopithecoids while being Old World monkeys, are so basal they share some New World monkey traits to the exclusion of both apes and Old World monkeys such as in hand morphology. Pliopithecoids lived from about 18 to 6 million years ago, last known at about the same time as hominins appear in the fossil record, which may not be a coincidence. Proximal femurs of pliopithecoids and early hominins are quite similar, though this is likely to be from convergent evolution. Pliopithecoids are rarely found together with great apes due to pliopithecoids preferring wetter environments than great apes, sometimes wet savanna type environments with mixed trees and open areas. Early great apes experimenting with more bipedal movement on open ground, in a similar manner to bonobos, might have found adaptations to this type of bipedal movement by earlier pliopithecoids useful through HGT or inter-specific introgression.

The dryopithecine apes are the best candidates for being ancestral to the great apes, and some of them were bipeds. The dryopithecine ape Danuvius (Bohme et al 2019) from the Miocene of Germany about 11.6 million years old is an exciting find. Danuvius is the first documentation of locomotion referred to extended limb clamboring, in which they stood upright holding on to branches. Danuvius had human like legs and orangutan like arms. Some humans today, including the author, use a similar type of locomotion crossing rugged terrain or moving through dense trees. One Danuvius fossil is ape like in the front of the upper jaws being relatively square, another is rounded like hominins. Danuvius suggests the initial morphological divergence between great apes and hominins was early, having some hominin traits not seen in other great apes fossil or living. "extended limb clamoring thus combining adaptations of bipeds and suspensory apes and providing evidence of the evolution of bipedalism and suspension climbing in the common ancestor of great apes and humans" (Bohme et al 2019). It no longer appears bipedalism began as an adaptation in human ancestry to living in treeless areas. It now appears bipedalism began in the trees. And once living on the ground, ancestral humans developed thicker leg bones to better bear long duration weight of the body without support of the arms in a standing position.

All known great apes have high sexual dimorphism compared to all known hominins, which are known for sexual dimorphism, except some Australopithecus cross the line. Gibbons are monogamous and do not live in groups except bonded pairs and their sub-adult children with low sexual dimorphism. Some apes, such as Gorilla, and it seems some Australopithecus based upon trackway evidence, live in bands with a dominant male and his wives and male subordinates. The Homo ?naledi tracks from Crete at 5.7mya suggest a similar family structure. The first morphological hominins may not have been hominins. They may have been diminutive great ape males, with a morphology adapted to sub-ordinating to larger males. Without males, it is problematical to distinguish early hominins from great apes, as the difference is mostly based upon lower sexual dimorphism and smaller canine teeth in males. Males were likely to have been morphological hominins before females were. A morphotype adapted to be non-threatening to alpha males as to avoid being driven out of the band (which usually leads to death). Females may have found the child like morphology attractive enough for the morphotype to be persistent in the population, with such males enjoying protection from their mothers throughout life as they had juvenile (neotonous) morphology stimulating protective behavior in females.

With the development of more advanced weapons, such as sharpened sticks (spears), alpha males would have found advantages to have a retainer of such diminutive males, who did not require as many calories to maintain as larger males. An alpha male with a cadre of spear wielding adult males, would have had great advantage over males who had driven all adult males from their band. Orangutans are solitary except when caring for young. Dominant males develop distinctive flanges around the face, and females show strong preference to mate with males with flanges over males without flanges. In orangutans morphological adult male dimorphism is based upon life experience, phenotypic expression expressing two types of males from the same genomes depending upon success in life. In orangutans, in identical twins, one male could have flanges and the other not. Many animal populations have large body size males and small body size males. Often the small males mate at night, and larger males mate at day time.

Big bad Jean is at advantage often in mating, but he is at higher risk for mortality. Big bad Jean's friend little John, can also be at advantage in mating through the protection and sharing of his friend Jean. And back door Jack, is maintained in the gene pool as well, by stealth mating when big bad Jean is sleeping or away. Though Jack can be at higher risk for injury or death, for example when little John catches him and calls out for Jean. Both alpha chimpanzees and Gorilla males allow their subordinates to mate, after they mate, as is a common pattern in human society. Charles Oxnard found that there were equal numbers of large and small individuals in Gigantopithecus in a tooth assemblage suggesting an even split of males and females. In Pan troglodytes (common chimpanzees), average body weight of males is only about 18 pounds more than females, some human populations have this high of sexual dimorphism. Gorilla has relatively high sexual dimorphism in terms of body size, which can be related to subordinate males usually mating when the female is infertile.

Gibbon males are only slightly larger than females, so hyper-human in this trait versus great apes. Orangutan males are about twice as heavy as orangutan females and have high variation in body weight among males "Alternative male reproductive strategies: male bimaturism in orangutans" (Utari Atmoko and van Hoof 2004). African living great apes are likely to be more derived relative to populations ancestral to both great apes and hominins than orangutans. Environment is influential on sexual dimorphism. Low food yield environments often will not support large bands living in one group, not enough food so they need to be spread out. Which is thought to be why gibbons are monogamous and with almost no sexual dimorphism, as they live in low food yield forests. When apes and hominins could challenge large predators, they needed to form closely bonded bands of larger size. Large predators often work together in groups. In archaic human times in Eurasia and Africa, there were packs of lion sized hyenas. In living humans males average about 10 percent larger than females, though in some populations the ratio is much higher.

Lufengpithecus is a strictly Asiatic great ape that is known from about 12.5 million years to 200,000 years with body weights of about fifty pounds. Lufengpithecus is known for post cranial material suggestive of considerable bipedal abilities. Lufengpithecus was of similar size to Homo floresiensis, and if juvenile craniums of Miocene Lufengpithecus are compared to the Homo floresiensis cranium, they are remarkably similar. Juvenile Lufengpithecus craniums are closer morphologically to Homo floresiensis that to any known African ape craniums. This could be a result of convergence it could also be a result of Lugfengpithecus early introgressions into hominins to yield the genus Homo. It is likely some unknown populations of Lufengpithecus obtained hominin grade. Lufengpithecus has been misidentified as Homo erectus for late wear state teeth.

Wolpoff 1982 suggested Ramapitheus (an Asiatic ape which at the time included Lufengpithecus) was ancestral to humans and all great apes. Lufengpithecus was first assigned to the genus Australopithecus and then to the genus Ramapithecus and finally named as three species in the genus Lufengpithecus. Grehan and Schwartz 2009 found that great apes and humans are monophyletic and compose two clades. Living African apes in the first clade and hominins and orangutans along with a various fossil great apes including Lufengpithecus in the second clade. The main protest to these types of classifications by Schwartz and his coauthors is that chimpanzees are closer to humans than orangutans in whole genomic distances. There is a simple solution to this seeming contradiction.

Orangutans could have become genetically isolated from other great apes and humans relatively early, such as by isolation on Asiatic Islands. With ancestral chimpanzees diverging before orangutans from ancestral hominins. With chimpanzees being in proximity of ancestral hominins and failing to speciate from them so early as orangutans. In addition, orangutans are less derived than chimpanzees so preserve relatively more of the morphology of populations ancestral to all great apes and hominins than chimpanzees. Fossil Asian great apes, which are unknown for genomes, might have become genetically isolated from ancestral hominins much later than their Western branch orangutans, perhaps even later than chimpanzees became genetically isolated from hominins. Moving East, Homo shows more Asiatic great ape characters even until today, than Europeans and Africans fossil and living. In Africa, hominins show more African great ape traits than are seen in Asia. Introgressions between great apes and hominins after initial divergences are well evidenced both in Africa and Asia and are the foundation of lasting differences between hominin populations geographically. Chimpanzees need not share more morphology related genes with humans than orangutans to be closer to humans whole genomically than orangutans. If early Homo was exposed to more diseases from chimpanzees than from orangutans, as is the case today, then it would be expected the immune system related part of the whole genome, much larger than the mall spart of the genome related to morphology, would be more chimpanzee than orangutan, under natural selection.

In Thailand Lufengpithecus living in a flora unbroken with the same flora in Africa, had essentially orangutan dental morphology, but not orangutan like bones. Suggesting Lufengpithecus was close to the unknown Miocene roots of orangutans. Lufengpithecus is suggested by the fossil record to be close in ancestry both to hominins and orangutans. Pierolapithecus appears close in ancestry to chimpanzees and Gorilla. Moving back in time to around 7 to 9 million years ago, all hominids (great apes and hominins), were probably the same biological species except ancestral orangutans and began to speciate from the rest one by one. Until leaving the four relic populations of today that are all speciated from each other. With more Western traits in the West of the species, persisting until today, and more Eastern traits in the Eastern range of the species, persisting until today.

Asiatic great apes and hominins look Asiatic just as Western ones look Western, through time. With hominins covering the full original range and retaining local gradational relic morphology of the original species. Differentially over distance roughly corresponding to the geographical origins of the variable traits in the original single hominid biological species. Genetics now suggests that when archaic human genomes are found, the relic morphology of those genomes in and around where they lived came exactly from those genomes. Not a result of convergence as was thought by nearly everyone in the West up until a few years ago. Genetics only compares known genes, which for great apes means only fully modern relic great apes which are different in proportions from their ancestral populations and different in their whole genomic distances over time.

The neutral hypothesis is now widely considered invalid in science. With more and more "neutral" genes being found to not be neutral at all. DNA with no advantage, or potential advantage in the future, can seem cheap enough to maintain unwanted and unused in the human genome. However, nature usually removes waste over time through natural selection, even if the burdens removed are not large. Molecular clocks rely upon the validity of the neutral theory. Without the neutral theory, molecular clocks are just at times good estimates, not actual "clocks". And sometimes when there is no fossil record, molecular clocks are the only estimate available. When there is a fossil record, and "molecular clocks" wildly disagree with the fossil record, then most likely the molecular clock is broken some way or has been incorrectly retrieved or interpreted. Interpretations of fossil records are not completely accurate, especially of difficult records as for hominins. However, there is mounting evidence interpretations of the fossil record are over all more accurate than molecular "clock" estimates and interpretations. Human genetic studies gained priority over Western paleoanthropology because Western paleoanthropology was deeply flawed in multiple ways.

CHAPTER 7

Taphonomy Paleoenvironmental Studies Trace Fossils and Sedimentary Geology

Study of human fossils includes much contextual information in addition to the information the fossils themselves yield. Such as age of the fossils, relative to other fossils and absolute. What sort of environments the fossils came to rest in, versus what type of environments they usually lived in. Sometimes they usually lived in the type of environment they came to rest. Often places where fossils are found are atypical, so the fossils places of rest are different from random samples evenly spaced over time and distance. Evidence on the fossils, and in their surroundings, of how humans died and what processes acted upon their remains after death. Fossils are all of people who were dead when fossilized. How are death assemblages different from typical life assemblages for any given fossil(s)? What other organisms lived in the times of fossils, and how humans interacted with those other organisms, exploited them, and were impacted by them. What the relative abundance of each type of organism is in the fossil record, and what the actual relative abundance was. What sort of environmental preferences the animals and plants had associated with the hominins. Climatic changes and trends suggested by changes in faunal (animal) and floral (plant) conditions over time.

Fossils have sedimentary context. What were the sources of the rock fossils are in. If in the base of a thick volcanic ash bed which was a single fall, they might have died as a direct result of the volcanic event. Or have been attracted to the rich environment created by ash falls within a few years. If the bed the fossil is contained is was a river (riparian environment), and the rocks were transported by flowing water before coming to rest as evidenced by wear, the larger the size of the rocks the faster moving the water was. If in a lacustrine (lake) bed, they might have floated out into the lake after they died, been carried there by predators or scavengers, or been drowned or killed in or next to the lake by predators such as crocodiles. One Homo habilis foot has a healed crocodile wound. Suggesting a possible rescue by companions, even an armed rescue. Dogs do not normally survive being wounded by mountain lions without help.

Knowing which way water was flowing when fossils came to rest, or which way was up and down hill, can assist in knowing what direction in the bed to excavate for more parts of the scattered disarticulated individual. Carnivores sometimes leaves species diagnostic marks, such as leopard bite marks on the skulls of hominins from being drug by leopards by the head. Hyenas have powerful jaws which can crush bones in diagnostic ways. The acids in cat stomachs digest bone much more than dog stomach acids do. Birds etch bone fragments in their stomachs in diagnostic ways. If a bone has been well chewed by rodents, that usually means the bone was on the surface for some time before burial. Often chew or bite marks can be attributed to a particular animal species, sometimes one that was not otherwise found at a location.

Fossils can be found in the same position where they first came to rest upon death. They can have been reworked, which means they at some point were removed from encasing sediment such as by water from where they came to rest initially, and were then redeposited in a different place, and often in a bed of younger age than the initial resting place. Or washed out on to the surface of an older bed. This appears to have been the case with the Sahelanthropus skull, so the skull could be older if worked up, or younger than the bed it came to rest on if washed on to the surface of an older bed. Fossils are found intact in the sediments they came to final rest in or have been washed out resting on the surfaces of different beds to be covered up again. Usually washed out surface remains are coming from younger down to older beds. This same process can have happened in the distant past, with the fossils reburied, creating the appearance the fossils are younger or older than in fact washed up or down in stratigraphic sequences on to the surfaces of unconformaties.

Different populations which did not live together in time or space, or not usually, can appear in the same place and / or time as fossils due to postmortem transportation bringing them together or bringing one together with the other. Fossils are often mixed as to environmental preferences by being transported away from their normal habitats and mixed with fossils from other habitats. Usually fossils are not preserved in organically very active deposits, so a false perception can be generated of taxa preferring to live in more sterile places. Animals preferring to live in more sterile places can be over represented in fossil assemblages, such as some types of reptiles. Parts of animals and types of animals which are more likely to become fossils, such as tortoises, animals which died in burrows, animals likely to be eaten by carnivores who bury their kills and do not always recover them, will often be commoner in the fossil record than in life time assemblages for regions.

Following the rules of superposition tell us the relative, but not absolute, ages of bedded rock and soil deposits. If not reworked, materials deposited on top are younger in succession. Fossils in the beds, if not reworked or laying upon unconformities, are of age very close to the age of the rocks containing them and follow the rules of superposition relative to other fossils. Learning what types of sediments, or which beds or parts of stratigraphic sequences, are more likely to yield fossils, assists in finding fossils in the field. Often fossiliferous beds yield no or very few fossils at the surface, the fossils decompose before reaching the surface such as by water percolation. Bedding plains are the contacts between different depositional events. Except in paleosols (fossil soils), sediments are not deposited gradually, they are deposited, and removed, episodically. Doing field work during heavy rains and floods gives insight into depositional and erosional patterns. Fossils are usually found on erosional surfaces, and normally created in depositional environments. There is a strong bias in the fossil record for taxa that preferred to live in depositional environments such as river valleys or cave floors, versus erosional environments such as on mountain sides.

Taphonomy is a relatively new study, of the processes acting upon organisms after death. Such as observation of animal remains over time after death. Taphonomic processes continue after fossils come to final rest. Burrowing animals, insect acticity. Giving way to diagenetic processes. Compression by the weight of overlying sediment bearing down. Chemical reactions. Water percolating through sediments removing water soluble compounds and / or depositing water-soluble compounds in supers-saturation. Replacing compounds with different ones. Heat volatizing off some compounds and altering others. When excavating fossils important information may not be obvious, such as impressions of dried skin or signs of mechanical disturbance of the sediments such as the animal struggling to free itself from mud. Bones and tracks are not normally found together, just as fossil leaves and bones are not normally found together except in shales.

Ichofossils, also called trace fossils, are fossils which are not actual remains of organisms, instead they are marks in sediments left by organisms. Often it is difficult to impossible to know if a feature is natural or was made by an organism(s). Marks left by dead people entering and leaving caves are a type of trace fossils, as are tracks and burrows. Archaeology is arguably a branch of ichnopaleontology. Studying marks left by hominins, such as on stones to make tools, wear from use on tools and upon wood to make charcoal.

Because a something is not present, or is not detected, does not mean it was not there. Field parties can collect vertebrate fossils in an area for years and find few to no small mammals birds or reptiles. Where others could find many, with better methods, more interest, or by removing likely to be fruitful sediments and washing them picking the concentrate for fossils. Often animals were living in a bed, and are not present as fossils, while other animals are present. For example, if all the fossils come from the scats of some type of carnivore(s), which only ate certain types and sizes of animals, only those animals are represented as fossils. If fossils are only from animals who died stuck in mud, only animals which would get stuck and die in that type of mud, will be present. Surfaces which were stable for long periods sometimes accumulated bones and teeth over time with those surfaces becoming enriched for fossils relative to inside of the beds. Proximal mud flows are sometimes enriched for fossils by the sediments being washed away leaving the fossils. Moving water pushing rocks along in the current grinds up the fossils, often to nothing or small rounded fragments of bone. It is common to find fossils which were fossils when they were redeposited, often the wear on them is diagnostic because "green" (fresh) bone acts differently than fossil bone to transportation.

If the water in the ground is too acid, bones and teeth will be etched away, and no fossils will be preserved in deposited sediments. Generating a strong taphonomic bias against hominins living in areas of acidic soils such as forests. Paleoanthropologists have often had incorrect ideas about timing of first habitation and regions of habitation of forests by hominins due to this bias. If rock under forest soils are limestones, then fossil hominins are sometimes preserved in caves in forested areas. The fossil record for both fully modern human pygmies / negritos, and archaic pygmies, is poor relative to their abundance in the past due to preference for living in dense forests. Which contributed to incorrect ideas about pygmies being a much younger morphotype than they are. When there are no clear progressions through time from one morphology to another, the origins of morphotypes are unknown both in time and space.

If a region does not have any sedimentary rocks of a certain time frame exposed at the surface, such as in heavily vegetated areas or areas dominated by rock types other than sedimentary, then it is very unlikely any hominin fossils of the unrepresented age in that area will be found. Areas with heavy vegetation, without rapid erosion exposing sedimentary rocks of the right ages, and without investigation for hominin fossils in rocks of the right age exposed by Earth moving operations, will not likely yield hominin fossils. The amount of energy, money, and expertise spent to find fossils generates biases for certain regions and times. There is a strong bias for human fossils in Europe. With the success of Leakey, there has been a strong bias for human fossils from first East Africa and now all of Africa. In the times when it was thought humans originated in Asia great efforts were expended in Asia generating biases for Asia. China and Israel now have relatively strong paleoanthropology programs, generating biases for China and Israel relative to the rest of Asia. There is a need to expend more energy time and expertise in warmer parts of Asia. In recent years there has been a shift towards finding important human fossils in caves outside of Europe.

That hominin populations are not known by fossils, does not eliminate their influences upon other hominin populations such as by ancestry or introgressions. Seeing the fossil record of hominins as a sort of nearly complete temporal and geographic assemblage, with only a few missing parts, leads to inaccurate perceptions. Places which yield human fossils are not cradles of humanity and were not preferred high population density hominin environments. They are places where human fossils were relatively likely to be created, have since come to be exposed on the surface or locatable some other way, and are where people are actively looking for such fossils with the understanding of how to recover and publish them. Many hominin fossils in China for example, have been ground into powder for medicinal purposes or sold to wealthy collectors who do not publish them. Many hominin fossils are destroyed during construction and road building, without any awareness they were there. Paleoanthropology is a luxury some societies indulge in more than others.

CHAPTER 8

Lithics the Americas Beringia and Sahul

Wherever humans are living, traits appear which impart advantages in natural selection, and those traits are amplified and spread out until they reach places where they are disfavored by natural selection. The study of human morphology and adaptation begins with monkeys and apes up into the first unknown hominins interbreeding with great apes, and has no end unless hominins become extinct.

The manufacture of worked stone tools began in Africa at about 2.6 million years ago or before. There are claims for older stone tools which are not widely accepted and may be geofacts (made by nature not hominins or other animals). That capuchin monkeys make "Homo grade" stone tools (flakes struck from cores Proffitt et al 2011)), was not learned by Western science until recently. It is likely crude stone tools exist older than 2.6 million years ago, made by hominins and / or other primates, and are not yet detected or not yet dated accurately. Stone tools may be known from SW Asia as early as 2.5 million years. That stone tools are searched for, but not found, in sediments does not mean there were no hominins in the time and place the sediments were deposits. There might have been hominins there which did not manufacture stone tools but made tools from other materials not preserved, made or used such crude stone tools they are unlikely to be recognized as stone tools, or only rarely made stone tools. Bone and especially wood tools often will not survive the passage of many thousands of years.

Sharpened sticks / spears, such as some chimpanzees make and use to hunt, are unlikely to be preserved or found if of much age. Savanna chimpanzees have been reported sharpening sticks with their teeth and using the spears to kill a bush baby. With juvenile males and females hunting with spears, not adult males (Pruetz and Bertolani 2007). Western African chimpanzees independently invented stone tool use "possibly as recently as 200,000 to 150,000 years ago." Michael Haslam 2014.

The continent of Sahul during colder periods of the ice age included New Guinea Australia and Tasmania. From the beginning of the fossil record for hominins in Sahul, until today, there has been a mix of quite archaic (archaic Homo sapiens like and Herto like individuals and populations), and quite modern appearing humans. Fossil humans are known from Sahul only back to about 50,000 years. Adcock et al 2001 reported the recovery of an early Original Australian mitochondrial DNA deeply divergent from all other known hominin mt DNA. Which they suggested was not of African origin. With no linkage to fossils, there is no way to reliably guess the geographic origins of fossil DNA, where it is found often will not be where it originated. These results were hotly disputed for two reasons. The first is that many interpreted the claim, though the authors did not interpret it this way, to mean that early very modern humans in Australia were not closely related to living Original Australians. This is incorrect, as mt DNA is only a tiny fraction of the whole genome and does not usually indicate the degree of relatedness of different very ancient people. MT DNA us only from the maternal lineage, which far enough back, often will have contributed only a tiny part of the whole genome or appearance of people. The second reason this finding was hotly disputed, and attacked multiple times with studies and publications, is the finding violated the precepts of the Recent Out of Africa hypothesis (ROoA).

Since the time of the Adcock et al 2001 paper, the ROoA hypothesis has been internally amended. Since about 2014, ROoA has fragmented into different hypothesis, all of which accept a small amount of Eurasian archaic genes are present in all non-African people today. A stray very early diverging mt DNA in Australia is no longer a problem for amended ROoA hypothesis. It is common that some parts of fossil skeletons yield DNA from the fossil individual, and others do not. That the attempt to repeat the results failed, does not imply they were invalid. Until the deeply divergent mt DNA is found again, the results are uncertain same as with most conclusions from human genetics for more than 30,000 years ago. By about 50,000 years ago Sahulians had polished hafted stone blades. The author was taught in school such tools arrived only with the Neolithic which began in Asia about 10,200 years ago and spread out from there. The second oldest polished hafted stone blades are from Japan around 20,000 years later. Such tools are very useful for making high quality sea-worthy boats and suggest around 50,000 years ago Sahul would have had net gene flow out of Sahul.

During cold periods Siberia and North America have been connected by a land known as Beringia. At no less than 24,000 years ago people were living in Beringia and leaving Beringia for both Siberia and North America. The Clovis First hypothesis dominated English speaking archaeology in North America for decades and persists in amended forms today. Asserting that no humans were present in the Americas before about 13,000 years ago. Identified by distinctive Clovis points, which are older from the region of Ohio and spread from there to Alaska. People who entered the Americas would not likely have never returned to where they came from, but that is a common conclusion, that people only enter the Americas and Sahul, they never leave, because such peoples are "primitive" or technologically regressed. In the last ten thousand years, due to large animal husbandry in Europe Asia and Africa, there has been disease-based barrier to gene flow out of Australia and the Americas. The Tuvan, incorrectly identified as basal Asiatic Native North Americans existing in Asia before more than 12,000 years, made it to Asia from North America about 800 years ago to become Turkic people so were able to survive exotic diseases for North Americans in Asia.

A strong Sahulian genetic and morphological signal is now detected in South America, at well before 13,000 years ago. The signal it appears was diluted by a large net gene flow coming from NE Asia to the Americas after 13,000 years ago. It is not known how people got from Sahul to the Americas so long ago, there is no evidence for such a migration in NW Asia. The Solutrean hypothesis suggests Europeans came to North America living on the sea ice connecting North America with Europe around 22,000 years ago as sealers. Some have rejected this hypothesis on the basis it is racist, claiming North America for European "white people."

The oldest known Solutreans are 23.5 thousand years old, people were certainly in North America by 24,000 years ago. Solutreans were not modern white Europeans, they were more closely related to Siberians of their day than modern Europeans. Nor is there any good reason to think Solutreans came from Europe, they could have come from North America, or more likely were mixed from Native Americans (and looked quite a lot like some skulls from Brazil), and ancient Europeans. The best fossil record for human skulls at more than 10,000 years now from North America is Mexico. The skulls, up to about 14,000 years old, are quite cosmopolitan and like Australian skulls a mix of very modern and very archaic appearing individuals. With Chimalhuacan man, at 12,000 years, appearing to be an archaic human if found in the Old World. With morphologies which appear NE Asian, Sahulian, and ancient European. In long lasting cosmopolitan morphological mixes, it is to be expected whole genomic distances will suggest a single genetic clade (people), through kinship relationships.

Morphotypes of different founding populations are often maintained through natural selection founder effects and culture. Human morphologists with interest in the Americas have understood for over a century there is a "negroid" aspect to Native American morphology. Skoglund et al 2005

Nature "Genetic evidence for two founding populations of the Americas" evidences the "negroid" morphological signal is Sahulian not African. No recent (which did not enter other continents than Africa or the Americas first), African genes have been detected in Native Americans before 1492. Sahulians and African people today have similar morphology, with very high whole genomic distances from each other. Sahulian populations arguably have similar genetic diversity to sub-Saharan African populations of similar size.

CHAPTER 9

Controversy and Perceptual Changes over Time

This author follows a long tradition of being critical of the work of other authors in paleoanthropology. As a young man and student of anthropology in 1980, the author could see that Native American people retained some morphology from their neandertal ancestors. Tim White informed the author that Loring Brace, a professor at UCSB where the author later attended, was of the same opinion. By the 1990's only a very few professional paleoanthrologists were of this opinion. Stringer in 1997 published that with certainty nobody in the World today has any neandertal ancestry, an opinion which was confirmed in publication once the first neandertal genes were published. Loring Brace, who is well known for being harshly critical of the work of his peers (and is read by few even today though he has been proven more correct than his peers), retorted that human geneticists were in error in their interpretation of the neandertal genome having zero contribution to people today. Adcock pointed out that the use of mt DNA only does not rule out introgressions. David Reich was hired on with his team to study neandertal genetics, a firm believer that people today have zero neandertal ancestry, and soon proved in fact people today do have neandertal ancestry, outside of Africa.

It is now widely accepted all people in the World today have some level of derived neandertal genes in their genome, perhaps a minimum of two percent. Neandertal genetic heritage such as in genetic testing services, is detected level, actual level is at least about two percent higher as a background neandertal level in all humans today. Correct terminology is "detected" levels when referring to some ancestry, as such levels are not actual levels. Identical triplets have given quite different results as to ancestry in genetic ancestry tests, which are not accurate below levels of about 15%. Such testing is based upon the methods of professional human geneticists, whose methods are likely to fail detect levels below 3%. For example, all people in the World cold have about 3 percent denisovan DNA today with failure to detect. For archaic human and Homo erectus genes in people today except neandertal and denisovan, there is no detection, because there are no reference genomes. Except for trace levels of HGT genes and mutations, all people today are mixes, in varying amounts of each donor type, of archaic humans, Homo erectus, other hominins and introgressive great ape genes.

Opinions can shift quickly among professionals, even within the span of careers of great icons, which can be painful for them. Change is often met with stubborn resistance, though the magic of human genetics, despite ongoing great inaccuracies in interpretations, has subjugated the mainstream of Western paleoanthropology. In the past the Leakey's were the great Western celebrities, as paleoanthropologists and field workers. Today nobody has such high status as they did. Several human geneticists, despite their ongoing high error rates and nearly complete lack of knowledge about below fully modern humans except two highly specialized relatively low population archaic Northern Eurasian populations, have iconic status and are pursued by the press. Now the greatest discoveries are only made in the lab.

Human change and diversity over deep time, remains the nearly exclusive domain of human paleontology even if that is no longer the public perspective thanks to the press and media. Each time human geneticists cast down iconic pillars of paleoanthropological wisdom, the findings are presented as if a new discovery, despite someone some place usually already knew for many years such pillars were incorrect, and publicly so. "Proof" derived by the study of human fossils has been more reliable historically than interpretations of human genes. In the current environment, in the press, with the public, and even with many professionals, only interpretations of human genes can constitute "proof" and trump all past work of the same principles and presented as "new" discoveries. With both the public and most professionals ignorant about the past works contradicting prevailing views over turned by human genetics. Prevailing ideas often are as much of an issue of authority as evidence. Professional paleoanthropologists like Dr Loring Brace, who established by detailed morphological studies since the 1970's that his Western peers were on the wrong track about the single population from Africa being solely ancestral to all non-African people hypothesis, are forgotten by the press in favor of "new" findings in genetics repeating their works conclusions.

The study of archaic human traits in living and historical humans is not racist. All people today in their morphology are mosaics of archaic human and Homo erectus morphology, with just a very few newly derived traits mixed in. The study of such traits is the study of people's morphological heritage. Denying that peoples sacred ancestors, whose bones lay under their feet, are their ancestors, and forbidding the morphological study of such heritage, is in fact an aspect of colonial racism and suppression of conquered peoples heritages. Many people today, including many tribal members, take pride in their archaic morphological traits. Especially in the last ten thousand years, the average human has become physically weaker and with on average smaller brains. We should be careful that calling someone an obviously fully modern human does not become a derogatory phrase. Or that the Geico cave men do not become a model for the false notion of inherent racial superiority. In many societies strength is admired, in combat, in ability to gather food, in apparent intelligence. Most people living in cities today are like children on rugged terrain and in close combat compared to neandertals. Perhaps we can accept humans, living and dead, even ones with ape sized brains, are all equally as well adapted to the situations their ancestors lived in. As we all have been in this World the same amount of time at any moment.

"human evolution in Pleistocene times evidently passed through a species of Australopithecus and then through Homo erectus and men of Neanderthaloid type to the modern varieties of Homo sapiens" (Elwyn L. Simons 1963). Paleoanthropologists of the 1960's considered themselves quite modern and progressive. Many younger professional paleoanthropologists today, can't fully grasp the meaning of this obsolete language from 1963 above. "Neanderthaloids" are today split by most into archaic Homo sapiens or "modern humans" such as Jebel Irhoud and their close relatives Hualongdong both of which lived about 300,000 years ago, and classical neandertals. Neandertals only now are coming to be accepted as having lived in China and are not usually accepted any more as having lived in Africa. What distinguishes people who founded neandertals in Eurasia and came from Africa, from neandertals, Jebel Irhoud like people, is mysterious today. There always has been at least some small level of genetic isolation between Eurasia and Africa. If one searches hard enough, with enough fossils, traits can be found, such as the triangular depression on the back of the skull, which can distinguish neandertals in Europe and some neandertals in China from all known African neanderthaloids.

Simons statement about Neanderthaloids is not as incorrect by today's standards, as it might seem, given his usage of the term. Many still might think he was correct about humans passing "through" an Australopithecus stage. This is unlikely to be correct on two counts. First, there is mounting evidence, that hominin ancestry does not pass "through" any stage in a linear progression. Linear evolution is widely thought in life sciences a false concept now. Life does not evolve to become more and more like God until all life eventually looks like a Germanic medieval male king on a throne. Life adapts to changes in the environment, with all life at any given moment equally as adapted, as evidenced by the fact it is alive. Humans have always had high morphological diversity, when we first became humans we had high morphological diversity inherited from high great ape morphological diversity. It is likely early hominin diversity was fueled by introgressions and HGT from other than great ape primates as well.

A few professionals today doubt that all our ancestors were contained within the spectrum of Homo erectus morphology. All human populations today, in most of their members, have Homo erectus type lower bodies capable of long-distance walking and / or long-distance jogging. This was not the case in China as recently as 14,000 years ago, where multiple lower body remains from several locations are "below" Homo erectus grade and were not capable of long-distance walking. Most members of some local Native American populations will injure their feet if they walk long distances, though they are excellent at jogging, with flat feet. There is no evidence these people (typified by Red Deer Cave people, a type of archaic pygmy), or any other archaic humans or Homo erectus, went extinct. Species go extinct, over time populations (morphotypes) within species merge into each other and transform into new morphotypes adapting to changes in their environments and cultures.

So far as is known, all populations of Homo erectus and archaic humans are alive in people today. "Negative" neandertal genes in people today do not demonstrate neandertal inferiority. Such genes would have already been taken out, except rare ones, by natural selection if they were net negative. Such neandertal genes associated with factors like ADHD and schizophrenia, imparted more advantages than disadvantages on our ancestor's genes, which is why we still have them today.

Simons "modern varieties of Homo sapiens" is a modern concept. In the 1990's it was widely thought there is a valid morphological definition distinguishing fully modern humans, Homo sapiens sapiens, or anatomically modern Homo sapiens (AMHS or AMHs). Wolpoff was severely criticized in print in 1994 for claiming fully modern humans only exist as a sort of fantasy, and not invited to present a paper at a paleoanthropology conference. This author has made public challenges for anyone to attempt to define what fully modern humans are by morphology, without a single taker yet. The ONLY morphological traits the author can find which defines all extant populations of humans from all archaic humans is the greatly enlarged Broca area with the corresponding impression in the cranium. Which trait appeared about 70,000 years ago some unknown place. Fossil human skulls are not normally broken open to see if they have the impression of the enlarged AMHS Broca area inside.

Naming chronospecies and subspecies of humans has failed under the rules of zoological nomenclature at every step from Homo habilis to Homo erectus to archaic humans to Homo sapiens to Homo sapiens sapiens. Which fact is intimately related to the complete break down of the widely accepted (until the last couple of years) concept of complete genetic isolation between different "species" of hominins. Paleoanthropologists are not usually strong when it comes to morphological variation in extant (living) humans. Loring Brace being an exception. At more than ten thousand years, all fossil records are vastly more impoverished for populations studies than with the studies of extant human morphology. If the fossil record were nearly complete, instead of appearing as isolated individuals and populations morphologically quite distinct from one another, they would gently grade into each by small clinal steps over both distance and time, with few exceptions.

Using Homo floresiensis as a model for unknown Eurasiatic populations introgressing with African fossil populations, each gradational step up from Homo erectus to AMHS is better understood as an introgressive step than evolution within a lineage. Increasing brain size being being led by the West (Africa and Europe), and pushing increases, and lately decreases, in brain size by genes such as ARHGAP11A. Prognathy reductions and more modern dentitions, by introgression from Asia. The ossified chin, Asiatic. Reduced or missing wisdom teeth, Asiatic. Now that each end of the human spectrum has a relic ape sized brain Homo, we can better see brain size is more environmental than phylogenetic.

Simon 1963 also writes that all new discoveries of the major stages in human phylogeny will come in the area of pre-Australopithecus. He was largely correct in this assertion. The discovery of relic ape sized brain Homo was quite unexpected by everyone, though it should have been anticipated as many already understood linear evolution is a false concept. That ape sized brain apes are alive today, suggests that ape sized brain humans also could have persisted until 2000 years ago in Palau, 75 thousand years ago on Flores, and 250 thousand years ago in Southern Africa.

Humans do not pass through stages. Instead a process of diverse experimentation in adaptation to ever changing circumstances, with some traits over time being relatively more favored than others. Very few new traits being created, and traits being passed back and forth between morphologically diverse grades of the same biological species. So ape sized brain relic Homo, are a "stage" humans passed through even after morphological Homo erectus was "extinct". Nor is there any wide agreement yet, if any hominin "stage" is actually known "below" Australopithecus. It may even be, that "below" the genus Homo, every "stage" is a mosaic, at least in some populations, of traits diagnostic for great apes and traits diagnostic for hominins, in different proportions in each fossil population.

In 2016 Haile-Selassie et al wrote a paper titled "The Pliocene hominin diversity conundrum: Do more fossils mean less clarity?" More is always better when it comes to interpretations of the fossil record. Though more will often challenge our ways of thinking. They also wrote "raised the possibility that early hominins were as speciose as later hominins". And that in the middle Pliocene of Africa there were other species than Australopithecus afarensis clearly distinguishable by their locomotor adaptation and diet.

Tim White in a 2017 lecture followed the standard doctrine of that time, stating that it was painful for great apes to stand upright and walk bipedally. Is this true? Certainly not, at least for some individuals for every type of great ape alive today. At the time it was already known Oliver, a Pan troglodytes, habitually walked bipedally, which was dismissed as some sort of aberration or surgical alteration (remarkably leaving no scars). This author disputed this false concept of ape locomotion. And fell under harsh public criticism for such ignorance of the literature and standard education. However, unaltered photographic evidence does not There is massive photographic evidence that all types of living great apes, in fact do sometimes comfortably and habitually walk bipedally, a benefit of social media. Osteology forms in young humans and apes to aid the types of locomotion of choice. The skeleton has considerable plasticity in young humans and great apes and shapes itself to suit the preferred types of locomotion. Humans do not knuckle walk today, but a recent Tarzan movie was correct in that Tarzans hands could form osteologically to suit knuckle walking if he knuckle walked from infancy. It was common professional knowledge only African apes knuckle walk today, again social media to the rescue, orangutans also knuckle walk. The author showed such photos on twitter to professionals, and now this knowledge has also spread from social media to paleoanthropology professionals by memes.

Only the most social media isolated professionals today might make the claim that great apes are obligate quadrupeds. Though some work remains to be done to convince all of them that humans also can be quadrupeds sometimes. It was widely accepted by professionals until a few years ago that great apes can't swim. This despite scientific reports that orangutans can swim long distances. Pet ape owners, of both orangutans and chimpanzees, found that apes when given access to water in fact do learn how to swim on their own. How did professionals go so wrong, proven so wrong by social media and films on you tube? This author was angrily corrected, until silencing the notion apes can't swim with you tube links. They had observed that zoo apes can be contained by moats. Captive animals do not usually like to stray far from where they are fed. Zoo apes do not want to swim, which is not the same as not being able to learn how to swim. Pet apes want to learn how to swim, and they do it on their own in a different manner than how their human adopted families swim. The author also likes to observe apes at zoos, however, today there are research resources in addition to field work, peer review literature and zoos. Now we have the luxury of social media and assistance in research by the public in mass.

Many paleoanthropology graduate students now are supplementing their education with social media. Social media allows like-minded people to congregate. Future paleoanthropoloists will be genetics and social media experts. The average peer reviewed paper in science gets about 15 readers including peer review. Most posts on social media science groups get a similar number of readers. Some paleoanthropology "bloggers" now, some professionals, others formally trained amateurs such as this author, others untrained amateurs (some of them well read), can get more than a thousand readers. The twitter posts the author sees, are mostly of a political nature with no scientific content. Mostly from professional scientists followed by the author, making twitter research more time consuming.

Ninety percent or more of professionals are ignored by nearly everyone. Social media frees them to be heard by a few dozen like-minded people. The big names get far less attention in social media than they do from the press, hundreds instead of many thousands even a million. Lee Berger and Tim White have publicly exchanged accusations that they are press first and science second in their discoveries. Is social media and the press a legitimate form of science and education? Many professionals point out that social media is a great engine for the propagation of false beliefs. Are they so sure all such beliefs are ALL false? To the point of scientific proof? Freedom of thought and diversity is a good thing, just as morphological diversity at any given time in a biological species is a good strategy to be able to adapt to change well. Professional human studies socially are already centered in social media, not peer review literature or conferences. Through social media, we become better informed about the literature than we can be by attending conferences and hanging around in university and museum labs. Great libraries remain extremely useful, but not everyone has one less than 15 minutes-drive or walk.

S. L. Washburn and R. L. Ciochon 1974. "If one looks back over the controversies of human evolution, they have one element in common, new discoveries, theories, methods came along which no one in the controversy anticipated. The "facts" changed,". Very predictive of the facts of the discovery of Homo floresiensis. Only it took some time for Homo floresiensis to evolve, in the literature, from some sort of freak or shrunken Homo erectus, to a highly functional and well adapted ape sized brain living at the same time as people with brains larger than the average today. Some still have the hobbit as severely challenged in locomotion, saved from competition by isolation. How would it come about people would devolve and lose the ability to move well, even on an Island?

In 2016 a fragmentary Homo floresiensis remains were found at Mata Menge on Flores Island (van den Bergh et al 2016), and as this author and others had suggested publicly on the face book record, they were smaller, and LESS Homo erectus like than the later Homo floresiensis fossils at around 75,000 years. This author endured considerable "schooling" in social media about how this view was no longer considered correct, and had been well disproven, and was behind the times. The next year in 20017 a paper was written suggesting Homo floresiensis was not a diminutive Homo erectus, instead diverging from a grade about at the level of Homo habilis or even a little below, and all hypothesis rival to this one fell nearly silent. The point is usually missed now, that Javanese Homo erectus traits in Homo floresiensis, are important, because they demonstrate that Homo floresiensis was a grade of the Homo erectus biological species, despite never "being" morphological Homo erectus, and was not genetically isolated from other hominin morphotypes in the region of Flores Island. The hobbit had access to bigger brains, and for a million years, did not want them.

False ideas still yield useful information, once the information gained by studies with false conclusions is correctly interpreted. Javanese Homo erectus traits in Homo floresiensis are mosaic, not well blended, suggesting large chunks of DNA moving together to make the traits. Not relic traits retained over a long time period. Dental morphology has underlying genetic mechanisms. Often these mechanisms act upon more than one tooth in each side of the jaws. The canine tooth in modern mammals is like a relic reptile like mammal morphology. Sometimes canine type morphology is seen spread to other teeth in modern mammals, such as in the premolars of Homo tsaichangensis. It may be canine like morphology is useful in gripping slippery raw clam meat with the teeth. When genes are not expressed, it does not mean they can't be expressed again in future generations in case they become useful again.

Traits can disappear, and then be revived atavistically, which complicates phylogenetic interpretations. Adaptive genes often are more about turning off or slowing down genes that have become negative from environmental changes, than doing anything new. Such suppression genes can easily get turned off in turn, such as by other suppression genes or being disabled by mutations. Many individuals today have archaic appearances, both in pathological and non-pathological individuals and populations. The archaic genes have been suppressed in the phenotype and are reactivated by mutations in genes suppressing the traits. Or by lucky recombination. In other cases, archaic looking skulls appear to have always existed since they were in fashion, in families or skipping generations in families.

Mr Valuev former heavy weight champion boxer, inherited his strongly expressed neandertal traits from his Tatar grandfather, they were not expressed in his parents. His grandfather like him, was famous for his great physical strength and endurance. Before it was widely accepted in the last few years in the West that people today sometimes express neandertal morphology, Mr Valuev was widely slandered and libeled by professional paleoanthropologists that his traits were from a thyroid disorder (he has publicly refuted these claims). Anyone with minor expertise on thyroid disorders knows just from photos Mr Valuev does not have thyroid disorders. Nor did these professionals seem aware professional television wrestling in America is largely acting, and not a competitive sport, so Andre the Giant, who does have a thyroid disorder, could wrestle despite his severe medical handicap of thyroidal giantism. Mr Valuev is now healthy and has healthy children, with none of the normal negative and severe by his age symptoms of thyroidal giantism. Mr Valuev is six and a half feet tall, a half a foot taller than known neandertals. Hybrids are often larger than either parent population.

Washburn and Ciochon 1974 continue, "challenge the concept of the one single-species. (they mean for hominins older than 3 million years old) Its proponents, however have raised a number of questions related to fossil species recognition based on small sample size and lack of a clear demonstration of ecological diversity to support multiple related hominim taxa.". In 1974 no likely hominins were known older than Australopithecus. There were two camps, one that there would be only one species and genus at more than 3 million years, the other that there would be multiple species in one genus. Against all predictions, four genera of likely early hominins have been named Graecopithecus, Ardipithecus, Sahelanthropus and Orrorin.

The diverse early known likely hominins, which expectedly would be only a small sample of the actual morphological diversity, need not be different biological species, especially given their wide geographical environmental and temporal range. They can be different grades of the same species, which included a mosaic complex of both ancestral great ape and ancestral hominin populations. Thinking of an adaptive radiation as speciating linear phylogenies, instead of an increasingly complex interacting gene pool, naturally suggests diminishing morphological diversity moving back in time to the "origin" of one "species" from another "species" which are instead grades in a species. The fossil record suggests that hominins did not branch out from a single point from great apes. Instead the generation of hominins from great apes was a complex interaction of diverse morphologies within great apes. With hominins not starting from a single great ape morphology, instead starting from diverse great ape morphologies to yield diverse mosaics of great ape and hominin trait bearing populations. Hominins emerging from a single biological species prevalent in the Miocene, with high morphological diversity over great distances, a single great ape biological species now named as several genera.

Geneticists in other fields than hominids (great ape and human), systematics have come up with results wildly contradicting the fossil evidence. The late Cretaceous and early Paleocene fossil mammal records are well known in North America, much better than for hominids. Following quotes from John Alroy Department of Paleobiology at the Smithsonian Institution. "several recent molecular analyses claim to show", "These claims confused basal splits with "radiations," employ exaggerated and unreliable molecular clock rates, and ignore the well-sampled late Cretaceous and Cenozoic North American fossil record." "a massive diversification took place during the early Paleocene,". Alroy continued "folly of denying the Paleocene radiation on the basis of loosely calibrated molecular clocks."

Western geneticists interpretations of data are under attack from Chinese professional human geneticists using the MGD method. MGD in this authors opinion yields results far more in agreement with the fossil record of hominins at more than 10,000 years and great apes than Western interpretations from human genetics, controlled by a small clique in Europe with control over the most advanced and cheapest sequencing methods in the World. And who generously have given free access to their genome data base to everyone. From Shi Huang 2009 "Early studies of molecular evolution revealed a correlation between genetic distance and time of species divergence. This observation provoked the molecular clock hypothesis, and in turn the 'Neutral Theory', which however remains an incomplete explanation since it predicts a constant mutation rate per generation whereas empirical evidence suggests a constant rate per year. Data inconsistent with the molecular clock hypothesis have steadily accumulated in recent years that show no correlation between genetic distance and time of divergence." In human genetics, whole genomic distances between human populations remain widely thought to correspond to timing of divergence (so to be accurate methods to construct family trees). Though more reliable methods appear to be emerging in the last few years with the many failures of such interpretations to construct accurate human phylogenies.

Drosophila (fruit fly) genetic researchers were among the first to rescue their field from these poor methods of interpreting data, as dictated by empirical results disproving the Neutral Theory. Human genetics and paleoanthropologists in the West have lagged other fields in discarding such methods. In mammalogy it is now widely understood that reliance upon the false neutral theory, can give false indications of migrations when there was only ordinary gene flow (movements of genes, acted upon by natural selection, versus movement of people). Some cultural anthropologists have long tried to explain that kinship relationships, are a better explanation for observed gene flow than full replacement, and now essential replacement, of Eurasians by Africans.

To compensate for flaws in false axioms, less parsimonious interpretations than others have been developed to support flawed models of reality. Especially convergence in place of shared traits from a common source, dead end lineages, and genetically isolated populations. The assumption that sub-Saharan Africa was genetically isolated from Eurasia, having living fossil populations such as the Khoe-San, has finally breathed its last only last year in the literature. Now human geneticists are scrambling to find new methods which do not use sub-Saharans as refence genomes in the false assumption that they are "pure", which has led to many incorrect interpretations of human genetic data. Without any archaic African genomes known, and no genomes known from Africa more than 8,000 years old, there are no reference samples to determine the geographic origins of genes in the sub-Sahara. Now that all resistance is ended this year 2020 to the fact that sub-Saharans are not pure African in ancestry, the actual geographic origin of the bulk of the fully modern human collective genome must be accepted as unknown in geographic origins since Homo erectus entered Eurasia at least 1.8 million years ago from Africa. Despite much effort, no evidence of even local extinction has come forth for any time period in Eurasia except lands flooded by seas or covered with ice.

CHAPTER 10

Human Cooperation with Animals

Primates are usually gregarious and cooperative. Orangutans are exceptional among primates in being solitary. Gibbons are monogamous and often live in pairs. Apes are relatively large primates, so require more food and are less vulnerable to predators other than large predators which are uncommon. For arboreal primates large enough size to deter avian predators, increases risks of falls and starvation. Leading to traveling between trees on the ground instead of the canopy. Environments such as some forests which do not yield large amounts of food, sometimes will favor small bands, pair bonded couples, or solitary individuals. Because the advantage of living in larger bands is outweighed by the disadvantage of a larger group not being able to find enough food for all their members.

Larger brain size in some populations was likely driven by cannibalism, the practice and avoidance of being cannibalized. With some of the very large brain and large body morphologies, likely starting as specialist predators upon other hominins and primates. Chimpanzee bands get most of their meat from other primates, especially young ones. Cannibalism is more of a strategy to clear competition than to acquire food. It is unlikely that at any given time all bands in a population are practicing cannibalism, as the risks imposed by diseases are high.

No primates except hominins are known to have subsisted from hunting large animals. Pygmies traditionally hunted elephants and Homo floresiensis did as well (Stegadons). In many Pleistocene regions of Eurasia fossil mammal assemblages do not appear to have included hominins, and large mammals were taken exclusively, or nearly so, by large carnivores not hominins. For hominins to move from gathering, folivery (leaf eating), frugivory, insectivory, seed eating and small game, as was probably the case for most Homo erectus, to a large part of the diet coming from large game, was a great hominin revolution which was new to primates. Requiring great advances in culture and tool making. Which in turn led to morphological and cultural adaptations to the new lifestyle. In some environments, such as many parts of the Western United States for Native Americans, there is not enough big game to rely upon it very much for food. Not only has human morphology always been diverse, the environments humans have inhabited have always been diverse. No one strategy for food gathering has ever dominated all members of the genus Homo. The upper Miocene was a time of great environmental instability, and severe competition from monkeys.

The author when living on his first remote mountainous orchard, observed his pack of dogs cooperating with coyotes and skunks. Subadult skunks and dogs will sometimes form lifelong alliances. There have been many documented alliances between animals in areas which are not forested, such as the Serengeti of Africa, where it is easy to observe animals from distances. In forest it is difficult to make detailed observations of complex animal behavior. When dogs become aware an allied family of skunks is coming for a visit, they become excited in a similar manner to when they see human allies that have been away. Sometimes dogs will leap through the air striking skunks with their paws in play with some scent released. Dogs will go out foraging with skunks for many hours, returning with a light skunk scent on them. Mountain lions and bears often eat dogs and wolves, having skunks in the pack helps to defend against such attacks. Foxes, mountain lions (cougars), coyotes and wolves have been known to prey on skunks, alliances with dogs can protect skunks from these types of predation. When wild animals are seen in cooperation, they usually already know and trust each other as individuals.

Coyotes which are a type of wild dog from the Americas, frequently form alliances with both humans and domestic dogs. The authors dogs would sometimes pack up with coyotes to plunder mountain lion kills of deer. From observation of sounds made at night by the mountain lion, coyotes and dogs a rough idea of the behavior was discernable. In turns the dogs and coyotes would drive off the mountain lion, eat some of the kill and dislodge parts of the kill to carry off, and be driven off by the mountain lion. The cycle would repeat several times in one night, and by morning only widely scattered bones were left. Predation does not always leave tooth marks on the bones. Dogs gnaw bones and also hold meat with their teeth pulling it free leaving no tooth marks.

When the first hominins, or great apes, cooperated with dogs (wolves, coyotes, jackals, wild dogs, domestic dogs, all the same biological species as it has turned out), which taxa was more experienced at cooperation in hunting with other species? Did humans domesticate dogs, or dogs domesticate humans and teach humans how to hunt big game? Some Native American tribes had domestic dogs which were morphological coyotes and mostly coyotes genetically. Coyote dogs can be quite friendly and cooperative but have a reputation for biting if one tries to touch them. In 1981 in the field, the author had a dream in which two coyote brothers were talking to him in English. They said it is strange that people think coyotes are not just a type of wolf. The author awoke to a pair of young adult male coyotes near him. And walked together with them several miles, the coyotes making some small animal kills. One brother would often come within 5 feet the other was more cautious.

Sometimes coyotes seeking to join packs of dogs will come near humans, often resulting in their death. Coyotes are not dangerous to humans, unlike dogs, if no attempt is made to touch them. Coyotes in populated areas sometimes become specialists in eating domestic dogs and cats. In remote areas it is rare for coyotes to prey upon domestic dogs, though rural Americans commonly believe it is common for coyotes to prey on domestic dogs in rural areas. When asked, they have never seen it happen, but attribute lost dogs to coyotes luring them away, such as with a female in heat, to kill and eat them. Dogs sometimes join packs of coyotes permanently, and usually do not survive long as they usually are not well adapted to sustained pack life. Rarely dogs are seen in packs of coyotes which are well adapted to pack life. The author over many years has had 10 dogs that would run with wild coyotes, often for many miles, and never lost a dog to coyotes or had a dog injured by coyotes.

Ancient peoples had many relationships with wild animals that are uncommon or unknown today or limited only to certain tribes or individuals. Amazonians by keeping pet monkeys, likely build resistance to monkey borne diseases over time. The monkeys also act as sentinels. Rarely humans today form alliances with Grizzly bears. Young male grizzlies even when allies, are dangerous to humans, and there have been a few recorded deaths of young men living with grizzlies or keeping pet grizzlies by young male grizzlies. The author was pinned by a young male rogue Grizzly when he was 18 to the South of Yellowstone Lake but not harmed. The bear had eaten a man a week before. Once in Montana a female Grizzly followed the author for four days in a sort of alliance, in which the Grizzly never allowed the author to see the bear. Predators moving together at distance from each other can benefit by game being startled and fleeing in the direction of companion predators. Projectile weapons, especially bullets, have greatly altered the relationships of humans to other animals. In the US and nations where most people have food shortages, animals of all sorts are at risk from firearms and other projectile weapons.

Most think early humans could not benefit from alliances with extremely dangerous carnivores. In the Pleistocene, many parts of Eurasia were effectively closed to human occupation, despite humans already had fire dogs and throwing spears, by competition with large carnivores. The

Romans had some understanding of ancient human relationships with large carnivores. Romulus and Remus were adopted by a she-wolf as infants. The Romans might have been aware such a relationship sometimes happened. In Aesop's fables an escaped male slave formed an alliance with a male lion, both actively hunted by the Romans and living in a lion's den. There are cases known of children being abandoned to domestic dogs and subsidized with food but not cared for by humans. In one such case the child was reported to be healthy and unable to speak except dog sounds (the parents were criminally convicted for neglect). Despite many rumors and legends, it is difficult to establish the existence of non-tribal feral humans today. Some tribes which have been severely disturbed socially by colonialism turn their infant children out of the home as infants to forage and rely upon charity without parental care.

Ancient people ate their dogs on a regular basis. The dogs were favored by their alliance with humans. Very dangerous large carnivores sometimes eat their human allies, this does not mean no ancient humans had alliances with large carnivores, enabling them to break into environments otherwise closed to humans. If archaic humans and Homo erectus had types of alliances not known today, or rarely known, or not known by western science but known by other cultures, why do those relationships not still exist or only rarely so?

As weapons became more advanced, the high cost of having a large bear, or pack of giant hyenas, as cohorts was no longer justified. The use of elephants gave a clear advantage in battle until canon. Elephants were a symbol of military power in ancient to medieval India. Today large carnivores sometimes seek alliances with humans, but they are unlikely to find a human companion both wanting such a relationship and having the ability to sustain such a relationship. In carnivore packs, weaker members are sometimes protected by alpha males and / or females. Humans have been under selective pressures to avoid being attacked as carnivore pack members, and to endear themselves to alpha males and females for their protection. Humans poorly adapted to carnivore attacks, can be frequently attacked by rural dogs, where other individuals are never attacked by the same dogs with the same relationship with the dogs. The most important adaptation to not being attacked by carnivores is not sweating adrenaline. Poorly adapted individuals get into negative feed-back loops, their past experiences driving fear, their fear driving carnivore attacks.

Humans, especially many children, have a strong instinct to adopt baby animals of all sorts. This instinct is so strong because it has at times imparted advantage. American wolves are not dangerous to humans, European wolves are. A little girl who has adopted or been adopted by a large wolf, is less likely to fall victim to attacks by wolves or other predators when gathering food in the woods. Carnivores with human companions, are less likely to be attacked by other humans. If the cost of what the carnivore eats and keeping it is justified by the services the carnivore gives plus the amount of food the carnivore brings in, then the relationship is profitable. Only in recent times did humans change the morphology of wolves by selective breeding. Before that, wolf morphology changed into dog morphology by natural selection. Shorter jaws, as big jaws were not as needed with human weapons in play to make the kill. Tails upright, to reduce the number of injuries and deaths from human projectiles, as the hunters could often see the tail wagging above the brush and grass. Better noses and higher running speeds less fighting ability and bite strength. Dogs were by natural selection developed to carry very few diseases and parasites relative to other carnivores that afflicted their human partners. Young men driven from their bands like young male canids often were without territory. Most dying when entering hunting grounds not occupied by humans, over time some surviving, and recruiting females.

Domestic cats are a recent adaptation with many obvious benefits to agriculturalists. If Reindeer herders of today did not have metal tools, there would be no agreed upon way in the archaeological record to distinguish them from reindeer hunters. There are traditions in the Americas and other places in which children makes pets of baby animals, care for them until they get bigger, and adults kill and eat the animals against the wishes of the children. Early animal husbandry was potentially a type of precocious maternalism / paternalism. The first people who protected their prey from other predators, and did not kill their prey at every opportunity, were the first people to practice animal husbandry. Homo erectus in Japan is now documented to have set posts in the ground to build huts. When posts were first set in the ground to contain animals is not known. In Africa the Maasai enclose their cattle at night with piles of thorn bush, something any people with stone chopping blades might have done.

Archaic humans, and probably the first Homo erectus to adapt to extreme cold in Asia around 600,000 years ago, often lived with their dogs. For protection, warmth, child-care and protection and companionship. The ability to control fire, and close cooperation with canids, were what allowed humans to newly adapt to extreme cold environments. Dogs often save their human allies lives from cold in North America today. The author has had personal reports from dog owners saying their dogs laying on top of them when unconscious saved their lives from cold. Dogs are often entrusted to care for children. Dogs often care for domestic animals without human presence, especially sheep and smaller sized animals which are more vulnerable to predation. The authors Border Collie will ceaselessly herd and defend chickens or rabbits on his wife's remote farm in Idaho where there are eagles foxes badgers and coyotes and near-by wolves wolverine and Grizzly bear. Rabbits are useful in places of extreme winter because they do not require nearly as much over winter feeding as larger animals with lower reproductive rates. Some neandertal populations ate rabbit.

Large animal husbandry greatly reduced the demand for and need for small animal husbandry, especially in the industrial age. San hunters say they know how to care for their wild animals, and they have many animals on their land. Modern industrial society concepts of animal ownership generate cultural biases confusing what the nature of animal husbandry, versus hunting. Compounded by lingering linear evolution notions demanding that archaic humans should have in all ways been less skilled at everything compared to people today, when they instead in many ways were more adept than people today.

Peoples occupations are what they are most skillful at and best adapted to. Modern industrial humans are well adapted to working in factories in polluted crowded cities, at inventing making and using industrial age weapons, not at complex relationships with animals they have often never seen except maybe in a zoo or are extinct. Just as humans have converged upon a single subspecies through purative natural selection and accelerated gene flow rates, domestic dogs have been reduced to a single subspecies within the last few hundred years. Dogs did not adapt from a single population of wolves and diversify morphologically over time. Dogs came from many different types of wolves jackals and wild dogs and sent genetic signals back into parent populations. The same errors in interpreting human genetic data have led to the same incorrect interpretations of domestic dogs. With derived domestic dog genes being interpreted as plesiomorphs (from common ancestors instead of from dogs).

In Soviet breeding experiments on foxes, two strains were developed from a single founding population. One strain instinctively friendly to humans, the other strain instinctively aggressive to and fearful of humans. Wild dogs until now can become domestic dogs, and domestic dogs can become wild dogs. Ancient Celtic people had free roaming pet foxes. Native American tribes of the deserts of Northern Mexico, who were never conquered by Europeans or colonials, form diverse alliances with animals. Such as sharing information by vocal signals with ravens. Some have free roaming chimpanzees since the 1990's. Ancient Celtic people had relationships with uncaged ravens, little is known about the details of these relationships. There are legends of ravens being used as spies. Some tribes of sea people are known to have had alliances with dolphins, and dolphins are depicted on ancient Greek coins with human riders. Assumptions that alliances with animals reported by ancient peoples, that do not exist or are unreported scientifically today, never existed, will often be incorrect assumptions.

Humans did not "conquer" environments they had previously been unable to live in without animal allies experienced at local conditions. They had animal allies that enabled them, along with their tools, culture, and control of fire, to enter new environments. Humans did not come down out of the trees to conquer nature. Humans were always out of the trees in some populations of great apes and expanded their ability through diverse innovations and new types of alliances with animals. In the Miocene frequent deforestation events dictated that apes could live without trees in some populations. Monkeys scramble for cover in trees upon hearing warning calls from birds. Diverse primates run bipedally when crossing open ground. Bipedalism is not only not a hominin invention it is not an ape invention among primates. In great ape populations leading up to hominins, morphological adaptations took place to allow enhanced ability to live away from trees. Which meant new types of animals, and new types of alliances with those animals, as well as old types of alliances.

In the upper Miocene local arboreal primate populations often went extinct due to major deforestation events. Great apes more capable of living without trees, including living around rock faces which they scrambled up (as suggested by Homo naledi foot morphology and local environment), were more likely to repopulate newly regenerated forests. As linear evolution is false, they did not need to retain their "superior" open terrain habits, they reverted to arboreal habits and morphology where they reentered arboreal environments. With less derived generalist morphologists and more derived morphologies of specialists. Every place great apes and hominins lived, they had great zoological knowledge and knowledge of animal habits in their favored environments. And they had diverse alliances with the faunas they were in contact with. Rabbits when they are very sick, have a death song, which every coyote knows by instinct. The coyote gets a meal, infection is removed from the rabbit population, cooperation. The dog remains with us, in part because they do not usually demand the lion's share which the human now takes.

CHAPTER 11

Gene Flow Rates, Captives or Founders?

Why would the differences between founding populations, and captive populations, matter in the human adaptive radiation? Before the use of metal, and before agriculture, societies did not usually generate enough food surplus to support nonproductive captives. Everyone in hunter-gatherer societies helps in some way, even if handicapped. Anyone who can see and make sounds can serve as a look out. The study of the ways things are now, or within historical memory, to deduce how things once were is useful. But can often lead to incorrect conclusions, as change is a constant in adaptation and the environment.

Humans are generally gregarious today, with a few exceptions such as people living in such low food yield environments living alone can be an advantage. Such as in deserts or inland arctic conditions. Great ape societies can be very abusive to some of their members, even lethally so. Captivity for more than a few minutes is not known in great apes. At some unknown time in human history, people started keeping other people captives, which over time became slavery. With strong consequences for human morphology and adaptation. Captive humans in many ways being like dogs once dogs began being selectively bred, in that their reproductive success became subject to the will of their masters.

Given the brutality and violence apparent in some archaic human populations, a good guess is that some took and held captives for at least short periods. Ropes can be chewed through or rubbed on rocks or wood until they can be broken. It is likely true slavery began with metal use. Only a few thousand years ago, so modern slavery has not had much time to impact human morphology.

Native Americans, in the Neolithic at the time of the first invasions from Spain, in a few cases in the copper age, did not practice hereditary slavery. In some Eastern American tribes, captives were either assimilated by becoming functional members of society, escaped or eaten by their adopted families. The Aztecs once captured a great warrior who was taken to their capital city and tied on a stone platform to a pole by a rope and given a weapon. He killed 15 warriors in 15 days before one killed him. Only the warrior who killed him, and his family, had the right to eat his flesh. Women visited the guarded warrior at night, in the hope of pregnancy. Modern slavery and conquest can generate a lot of confusion / cultural bias about patterns of archaic competition between tribes. Captivity in Neolithic people can be seen more as systems to avoid inbreeding, bring in useful genes, than as a means of economic profit or enablement of an idle wealthy class. Cannibalism was rare in the Americas and Africa before the 12th century AD.

People with neandertal morphology, in neandertal times as now, tend to be adept at combat of all types. Gilgamesh could not capture the wild man Eabani, who he desired as an ally due to his great physical prowess. Gilgamesh seduced the wild man with painted women to become his friend. Where the Y Adam lived nobody knows, nor is this very relevant to human morphology or where most of our genomes came from. It is a common instinct in humans to befriend or serve powerful warriors. Would slave keepers want neandertals for slaves, or enemies, or would they prefer more docile and gracile people for slaves or enemies? In modern warfare, war is on a mass scale and involves the production of a lot of metal weapons. Tribal warfare, without metal, is something that does not usually quickly result in conquest or territorial gain. Resources can be won. Hunting down every fleeing and hiding band of a tribe, killing all their warriors, and taking their people captives is not something that often will happen. Without metal armor, more advanced projectile weapons give advantage to the defender in rugged or forested terrain. Neolithic warfare, such a in New Guinea historically, usually was about balance between tribes not conquest. With reproductive rates keeping up with death in combat rates and territories not changing.

Founding populations recruit members of other populations and have kinship relationships with them. They alter morphologically over time in a quite different manner than "captive" people. Populations maintain their morphology through natural selection, not genetic isolation. So over time, they take on the whole genomic distances of their kin, not always their kin's morphology, unless there is advantage. Captive populations take on the male morphology of their captors. Do Slavic men look like sub-Saharan men? Do Jomon or Native North American men look like sub-Saharan men? That such long sequences of Denisovan genes are present in the region of Sahul today, suggests they were founded by Denisovans more than that they became captives of Africans.

Where European and African people essentially replaced Native / Original people in the industrial age, only in the Americas and Australia for continents, large amounts of metal were required to fill population lows fast enough for Native people to not refill them. There are medical studies which suggest, that after seven generations, originals / natives can become adapted to exotic diseases without introgressions with people from the Old World. Malinche was given to Cortez to appease him, as a slave. Aztec women obtained resistance to Old World diseases from European and African men quickly enough, that the Aztec people still exist and often speak their native language today. Industrial colonialism only accelerated gene flow, in 2000 years, the Americas will host Native Americans, and Europe will host Europeans. The populations will be morphologically and genetically distinct.

There is evidence modern humans ate neandertals, there is no evidence neandertals ate modern humans. It may be neandertals knew not to eat modern humans because they carried plasmid diseases, which can't be cooked out. Neandertal bones have been found in modern human graves. The normal condition is for people to kill each other at times and be at peace other times. In both conditions there is gene flow. For natural selection to act against a people being killed, they must be killed faster than they can reproduce. Before the iron age, it was very rare for people to be more likely to be killed by violence than disease. In times of environmental crisis, the leading cause of death is starvation and malnutrition. In environmental recovery periods after crisis, outside people can't move in fast enough to out race the reproductive rates of native people, to generate essential replacement. Not without gunships and resistance to animal borne diseases generated by some people living in filthy conditions for thousands of years with livestock. Most disease goes unseen in the human fossil record. For whole genomic distances, disease is the senior partner of combat.

Europe is much more studied for human genetics, archaeology and the human fossil record compared to all other places. Europe biologically is a part of Asia and less so a part of Africa, which are much larger places than Europe. Continuity for the entire length of the fossil record is evidenced in Europe. Europe compared to other parts of the World has had relatively high genetic and morphological turnover in humans, giving a false impression that such turnover rates are normal. The Kalash of mountainous Pakistan and Afghanistan are arguably one of the six whole genomic categories of all humans alive today (Ayub et al 2015). They are the most genetically isolated of any people, and the most internally exogamous with traditional Kalash marriage customs not allowing marriage if any ancestor is shared within 7 generations. Some Kalash are of Siberian appearance, others of Aryan appearance, and others of African appearance, with skin colors ranging from very white to black.

Aryans invading India did not need to take only non-Aryan women, there were already Aryan appearing women in India distantly related to the Kalash and more closely related to Iranian Aryans than to the Kalash. As Loring Brace wrote, skin color is not very important for determining human population types. Interpretations of human genetic data related to invasions of the last five thousand years are controversial within human genetic studies. Tracking ancestral relationships by whole genomic distances, Y lineages, mt DNA lineages, is often ineffective in getting accurate interpretations of data even at a few thousand years. Going back more than 14,000 years when there are no genomes for Africa or the Americas or any place except Europe and Northern Asia, is mostly wild speculation fueled by the press.

Seeing human morphological or genetic history as centered any place leads to incorrect conclusions. As the only archaic and oldest fully modern human genomes known are from cold parts of Eurasia. Centering the evidence the Arctic Ocean is the best choice. The Arctic Ocean has at times been navigable. This has yielded close relationships over time between the people of North America, Asia and Europe. There is no center in a gene pool or a globe, the edges are more relevant than the "center". Gene flow rates are faster along Coasts than across continents. And during environmental crisis, it is from the Coasts that there is strong net gene flow away from the Coasts in the period of environmental recovery. Centering genetic data where genes are found, is no more accurate than centering reconstructions where fossils are more often preserved and exposed on the surface today.

During temperature lows of the Pleistocene, the continental interiors, including Africa, were cold almost uninhabitable deserts. The human gene pool is like a whirlpools with gene flow rates and often gene densities much higher around the margins. There are inherently superior genes, such as those resistant to global pandemics. Placing a large amount of red dye in an ocean and tracking how it circulates until all oceans are red is analogous to such gene flow. The red dye does not fill all oceans, it becomes a very small detectable part of all oceans.

CHAPTER 12

Cryptic Apes and Humans and Oral and Written Traditions

The paleoanthropological literature contains little information about or scientific interpretation of historical sightings and oral / written of great apes and archaic humans unknown to science. There are large amounts of artistic renditions of potential historical archaic humans or unknown great apes. Ancient Hindu temple reliefs in Indonesia for some individuals in both body and facial morphology meet the authors expectations about how archaic pygmies might appear. The fossil record for "negritos" from Palau is about 2-3 thousand years old, for humans of unknown morphology in populations today, with considerable morphological affinities to Homo floresiensis. Homo luzonensis, also with affinities with Homo floresiensis, was first known from a single metatarsal (foot bone). And first published as the oldest pygmy known. The author pointed out on face book the bone had morphology much closer to Homo floresiensis than fully modern human pygmies, such as shared gradual distal tapering. More material was found, and the material was used to name a new species Homo luzonensis.

There is no evidence that archaic pygmies made or wore clothing. Some may have had clothes while others did not. Archaic pygmies as relic Homo habilis or even older hominins, suggest they might have sometimes had hairy bodies. Wide-spread and diverse Hindu legends which best match archaic pygmies morphologically refer to them as (translated to English) monkeys. Indonesian Hindu ancient temple panel carvings are of monkeys building a stone bridge thousands of years ago. Carrying stones in their hands to build the bridge. Some appear monkey like, others appear very human like, with full body hair, negrito like faces, and archaic pygmy type legs and locomotion. Though the reliefs are ancient, they would have copied older works of art potentially back to the art of people who saw the "monkeys" with their own eyes. In most languages apes are called big monkeys. These are the most credible and detailed renditions of likely archaic humans or cryptic apes the author knows about.

In the Central American jungles, the author has firsthand accounts of the tsetseguayape, a cryptic ape or hominin. There are also accounts in sensationalist magazines and other sources, and from cities in Honduras and Guatamala which the author has studied. The accounts from rural areas are of more interest and are often not reported to outsiders as missionaries have taught such beliefs are satanic and such creatures are demons. The tsetseguayape are often described as being the same as gorillas. Tarzan books magazines and movies have by now reached all parts of the World and reached central America by around 1920. Close up observers, when questioned for more details, do report differences from Gorilla. The females are reported to have large breasts (thought to be a hominin trait). They are said to often live in certain remote caves, not a known Gorilla habit. They are reported to wear ragged clothes, from people, and large hats which they weave themselves from leaves and use to carry fruit. In one report a family of three tsetseguayape were walking along a remote dirt road, a male female and child. There are reports of maderos killing tsetseguayape with 22 rifles. There are also reports of a little larger than known to science capuchins, sometimes smashing clams between two rocks. And feral chimpanzee bands. There is confirmed free ranging chimpanzee ownership in the region, with law-suits over damages caused by chimpanzees.

Tsetseguyape are reported to usually knuckle walk and sometimes walk bipedally such as when carrying fruit in their hats. In large areas of

Guatamala and Honduras tsetseguyape are common knowledge for at least 800 years. They are reported by some to have fire hearths, and sometimes live in very remote villages in stick and mud houses. The last Native American source said it was important to not look them directly in the eyes, which is a serious offense to male Gorilla (likely mixtures of stories about Gorilla and tsetseguyape, though he said he saw the villages himself). Tsetseguyape are said to frequent abandoned villages with standing groves of mangos and bananas, which were depopulated by the volcanic eruption of 85 years ago and can't be driven to. The author has seen both abandoned banana and mango trees with fruit that are not cared for, though in some reports banana trees die without care. Tracks are sometimes seen along rivers, and some are reported to live in nests in large dense foliage trees. There is ancient statuary of tsetseguyape, which is not realistic enough for the author to make any decisions about the likelihood of which are of cryptic apes or hominins and which are monkeys. There are ancient capuchin and other monkey figurines.

From Oregon there are stone heads published about 1900 reported as ancient. They appear to be authentic. The heads show great ape morphology. Up the Coast going North there is native art of cryptic apes or hominins, which are highly stylized though they do have some great ape morphology. It can't be ruled out that Native Americans had knowledge of Asiatic ape morphology by contact with Asia or remembered Asiatic ape morphology from when they were living in Asia.

A deceased friend of the author here in Utah reported seeing a female "bigfoot" while hunting in remote forested mountains of Utah. A female with breasts about five feet tall. A description much more like descriptions of tsetseguayape than "bigfoot". Some reports from mountains in Arizona also match descriptions of tsetseguayape much more closely than "bigfoot". Cryptic ape / hominin reports generally are without tails.

Bigfoot sighting numbers and geographical range of sightings have been steadily increasing in recent years. With very little in the way of physical evidence, if any such apes or hominins exist, they would be expected to be declining in numbers. Many cultures believe in spirits which can be manifested and interact with the material World. These are more likely to be the spirits of deceased beings than living beings, for most authentic sightings. British Columbia in Canada has enough food resources and dense undergrowth for the possibility of hosting undocumented by science great apes or hominins. The author has spent many years in remote Wilderness from Alaska to Central America and has never seen or heard any trace of a cryptic great ape or hominin. The author has seen only two wolverines, about six pine martins, and about six mountain lions. About is because often sightings can be so brief identity of the animal is uncertain. Bears at times can be rather bipedal and confused with bipedal primates. People have been known to wear outfits to fool people into thinking they are cryptic apes. A feral monkey lived in the area of the authors first orchard and was never sighted by the author.

Bonobos were cryptic apes for most of the life of the author. Giant chimpanzess or Bili apes are only in the last few years documented with convincing film evidence. Central America is the best place to look for cryptic ape / hominin remains or DNA (such as in nesting places in remote caves). An ancient tsetseguayape nest in a cave near Quimistan Honduras has been reported in the local press. The author took reports from the same area and from the same city where he has lived. The author has studied mammalian interchanges between North America and Asia, with years of field work and study. Without any recovery of primates, but with the understanding there have been interchanges of arboreal (forest dwelling) mammals through forested corridors between North America and Asia in the late Miocene and potentially in the Pliocene and in the Pleistocene.

Many Chinese professional scientists in different fields are quite convinced of the existence of cryptic apes / hominins in China, with similar situations in other parts of Asia. Large bodied intelligent animals can be very difficult to find and even harder to convincingly document when they have dense vegetation to escape through and hide in. In the jungles of

Central and South America, monkeys of all types do not allow the author to see them, except one time, while they often allow persons they know to see them. It can be hoped that new species of great apes living in historical times will be found and examined from living individuals or remains not killed by people.

A large book could be written just about oral and written histories of cryptic apes and hominins. Most Hindu art is long after the fact of any contacts and following long artistic traditions which are highly stylistic and unlikely to be based upon eye-witness artistic renditions. Hindu demons, and potential cryptic or reported hominin morphotypes, often are quite prognathic (jaws projecting out relatively to humans). Denisovans and Homo erectus would both be expected to have relative to extant humans high prognathy. There are many Hindu and Buddhist legends of monkey heros who can speak but do not know human manners and customs. They are often attributed with high martial abilities, sometimes are very badly behaved, and others are very well behaved. Sub-fossil cryptic apes or hominins could be very hard to find same as fossil hominins and great apes.

Many traditional stories have been thought to have been fantasies, and were later confirmed to be true, or partly so, by scientific evidence. The ebu gogo of Flores Island in many ways seem descriptive of Homo floresiensis. They could instead be much later archaic pygmies not known in the fossil record. Legends of little people are common in many parts of the World. Such legends could have traveled to Flores from places where little people existed and were observed, being altered over time. There are many different legends, with similar aspects. Such as the last of the little people being destroyed by fire in a cave by people. Stealing food human children or women. Reputations for cannibalism. Legends that could have traveled to Flores Island. The ebu gogo legends may be about relatives of Homo floresiensis, other grades of the same population. The author finds it unlikely anyone survived the Toba eruption on Flores Island of about 74,000 years ago, given the small size of the Island and proximity to the eruption. People using marine resources sheltering in deep caves might have survived. In some parts of Honduras and Guatamala only people living in deep caves survived the volcanic eruption of 85 years ago, the ash of which the author has observed and is recorded historically.

There are many legends in Europe of neandertal like people, often cannibals, many which go back to medieval times or the Dark Age. The Romans do not report such legends or barbarians having them. The Dark Age, around 500 to 800 AD, was terrible, and in following medieval times there were some very tough periods. People in those times did not perceive of their times being particularly bad, with some concept of greater past order. The legends are more often of individuals than populations. Grendel and his mother, both cannibals had the power of human speech. These sorts of stories could be generated in very difficult times by morphologies which exist today in Europe. Especially if a few cannibals were well fed while most people were small from poor diets. A human hero Gilgamesh, presumably well fed, was strong enough to defeat Grendell and tear his arm off.

There are a few published pygmies from Europe. From around 1900. While in scientific journals, the author has not found any that are pygmies, the illustrations are of individuals with dwarf, not pygmy, morphology. Pygmy populations are established for the Nile River by Roman art but are not yet established for Europe. An extinct pygmy tribe, smaller on average than any population of humans today, existed in Northern California in heavily vegetated areas. Ancient Oklahoma pygmies are said by Native people, before Columbus, to have liked very much to drum and dance. It appears likely pygmy morphology made the trip to the Americas. The author has observed pygmy morphology in Central America, which is either of recent African origins or more likely of ancient Asiatic origins. Pygmies are the first people known of Taiwan and some Formosan morphology today is strongly pygmy. Several Alaskan Native groups had clear contact with Asia such as the Tlingkit and the Inuit made regular trips between Siberia and Alaska before Columbus.

There are persistent rumors of red-haired giant legends in the SW of America. The author is not aware of any of these being from before 1800 AD. The Inuit have clear recollection of the Viking, which was discounted and has now been confirmed by archaeological evidence. The Aztec before Spanish contact had clear knowledge of Western European appearing people with sailing ships. There were probably Irish or Vikings.

Gigantopithecus are known only from jaws and teeth. Any conclusions about their morphology relative to other apes is highly speculative. There is a recent claim from proteonics that Gigantopithecus is established to be closer to orangutans than other great apes. The predicted date of divergence of Gigantopithecus from orangutans, at 12 million years ago, could as easily be the date of divergence (speciation) of humans from orangutans. Genes for dental morphology hardly can be expected to track whole genomes so deep in time or be accurate for phylogenetic reconstructions. Orangutan diets could have been like those of Gigantopithecus at some time enabling gene flow for dental traits between different populations, the donor being better adapted to the joint dietary habits than the recipient which had changed diet due to changes in the environment. Or humans could have once had the same protein and since lost the trait (a plesiomorphic condition).

A past director of the American Museum was convinced Gigantopithecus was directly ancestral to humans. The lingual dental groove is not up to modern standards to move Gigantopithecus from hominins to pongines (orangutans). The lingual groove for example could be a shared primitive (plesiomorphic) trait between orangutans and Gigantopithecus, not seen in hominins. Last year a group of young Turk professionals presented a chart at a paleoanthropology conference showing different great apes converging up into the hominins to form hominins. Strong mosaic traits do not so much suggest a convergence as a lack of full divergence / biological speciation.

There are credible, including scientific, reports from China of some large apes looking like orangutans. A Chinese scientific paper showed in photos what might appear to be Gorilla like great apes at very high elevation. Hard to make out, and it seems unlikely large primates could live at high altitude in Asia. Denisovans at least could in Tibet, but they had stone tools, dogs, clothing and fire. Many photos of cryptic apes and hominins are on social media. They generally are not clear or appear to be fraudulent. Captive Gorilla have been available for such photos and films for some time. It needs to be remembered that great apes living with people from infancy often move in very similar ways to humans, so can appear much more hominin like than wild great apes usually do.

Fragmentary remains are published from the Congo, less than 20,00 years old, with strong aquatic adaptations. Like Greek legends of water nymphs. Africa had at least one cryptic ape around two million years ago, which introgressed into bonobos (Kulwilm et al 2019).

If there are cryptic great apes and hominins, why are their bones not found and published? No bonobo bones were known before they were discovered. Hanno reported Gorilla in the 5th century BC, yet they were not known or believed in by Western science until Thomas Savage published some Gorilla bones and a skull in 1847. The mountain Gorilla was not accepted until 1902. Living great apes have been elusive even in the last 20 years in Africa. If even one new species of great ape could be found other than chimpanzee varieties, especially outside of Africa, it would be of great scientific importance. A new living hominin or intermediate hominin / great ape would be revolutionary.

AMHS have been very efficient at assimilating all other hominins into their gene pools, with no fossil archaic hominin morphotypes known by 10,000 years ago except on remote Oceania Palau. Any morphological relic archaic hominin existing now probably have already been discovered, are whole genomic fully modern humans, and are considered fully modern humans no matter how archaic they appear. Fully modern humans are defined by any morphology which has survived until today in the human gene pool. By definition any people alive today along with their morphology fully modern. Red Deer Cave people are the last known widely agreed upon archaic humans. If they were alive today they would widely be considered fully modern humans despite their very archaic lower bodies.

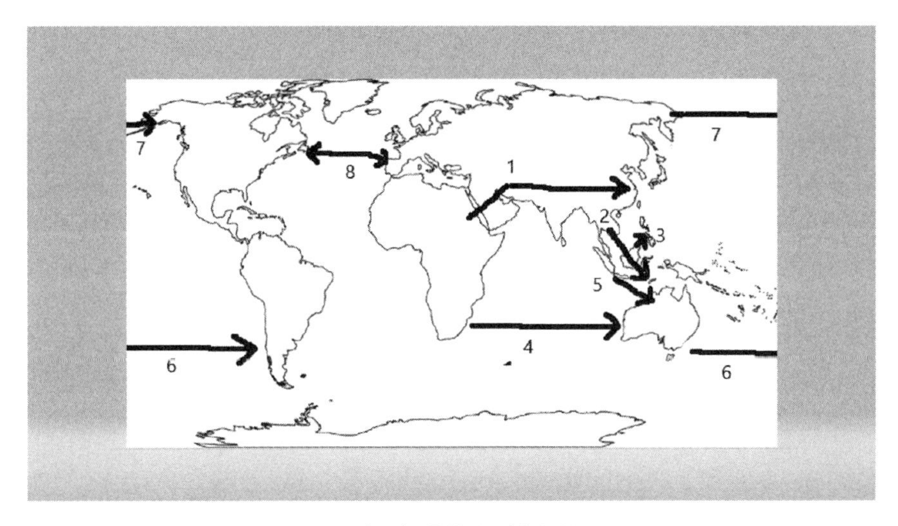

Map by Alan D VanArsdale 2020

Map 1 – Likely human continental interchanges estimated times of initial maximums 1. 2.1 million years 2. 1 million years 3. 740,000 years 4. 140,000 years 5. 80,000 years (two-way gene flow) 6. 25,000 years 7. 24,000 years 8. 23,500 years (two-way gene flow)

GLOSSARY

Adaptive radiation – Normally groups are restrained by natural selection to limited morphological variation for non pathological individuals. During an adaptive radiation, such as filling niche voids of other taxa left by an extinction event, greater morphological variation takes place which often leads to speciation. Populations fill new niches they have not filled before in adaptive radiations. Adaptive radiations begin with relatively less specialized and plesiomorphic (primitive) groups of populations.

Archaic human – A person who can be distinguished some way from any living or historical person, and who does not have an enlarged Broca area with a corresponding internal impression in the cranium. Individuals today can be essentially morphological archaic humans but could be distinguished from any archaic human genetically. Archaic humans should not fall within the norms of existing or historical populations morphologically. An archaic human must be a member of the genus Homo, and not be a member of the species Homo erectus or Homo habilis.

Archaic pygmy – A member of a fossil population of humans of the same or smaller average size of extant populations of pygmies and negritos. Archaic pygmies are not evidenced to have speciated from Homo erectus or Homo sapiens and are not within the morphological ranges of extant pygmies or negritos. Archaic pygmies have more affinities with Homo floresiensis and / or Red Deer Cave people than any extant populations of humans do.

Artifact – An object which has been modified by an organism for some functional, recreational, artistic or spiritual purpose.

Broca area – Part of the brain most related to learning and using speech in humans.

Chronospecies – A species which is defined by a time frame and not by traits which distinguish members of the species from all other organisms. For example Romans are not a species because they should include the descendents of Romans until today and all people before Roman times and their descendents who were more closely related to Romans than to any species not including Romans.

Clade – A group of organisms or populations which are united based upon shared traits. A clade may be monophyletic or polyphyletic. In cladistic analysis traits are weighted based upon how important the worker thinks those traits are as an indication of phylogenetic position. For example, feathers suggest membership in the bird clade today, but not in the Mesozoic.

Clinal - A gradation in one or more characteristics within a species or other taxon, especially between different populations

Convergent evolution – When two populations share traits which never existed in any common ancestor of those populations.

Derived trait – A trait which arose within the population it is seen in and did not arise in a common ancestor with the population which the population is being compared to.

Diagenesis – The study of the non-organic processes which act upon fossils (and rocks and sediments), after they come to final rest and are buried. The study ends for fossils once all traces of the fossil have been removed by diagenetic forces such as metamorphism.

Diagnostic – Fossils adequate to assign them reasonably to a taxa or population. Sometimes even with nearly complete material it can be hard to distinguish distantly related taxa from each other, such as with Artiodactyls in which teeth even skulls often are not diagnostic. Non diagnostic material still can be important.

Extinction – When the last member of a biological species dies.

Fantasy taxa – A grouping of organisms, such as a species, which is not defined in a way by which it can be distinguished from all organisms not a member of that taxa. Valid taxa should not have already been named under another valid name (then it is a synonym), should not as defined include members of other taxa, and as defined should include all members of the said taxa.

Fossil – Evidence organisms which are now dead existed. Fossil genes come from organisms which are selection.

Gene flow – The movement of genes through time and space acted upon by natural selection. So that genes which impart advantages are more likely to be spread than genes which are harmful given the current situation of the organism hosting the genes.

Geofact – An object which can appear to be an artifact, but which was made naturally without intentional modification by an organism.

Ghost, genetic – A relative to whole genomes small group of detected genes inherited, such as by introgression or founder effect, from an unknown archaic or other fossil population. Without attributing genetic ghosts to known fossils, there is no way to know their geographical origins as genes can change location over time.

Hominin – Any individual or population more closely related to any living person than to any living chimpanzee.

Hominid – Living great apes, humans, and their common ancestors. All living and fossil groups more closely related to living great apes and humans than to gibbons (Hylobatidae lesser apes).

Horizontal gene transfer (HGT) – The movement of genes other than by cell division or mating. Especially the movement of genes between species carried by microorganisms such as retrovirus.

Ichnofossil – Also known as trace fossils. A fossil which did not require the death of the individual that made it.

Introgression – When a population receives genes from another population by interbreeding with that population. Intra-specific same biological species inter-specific between different species. Intra-specific introgressions being common and between different species gene flow uncommon and rarely detected in fossils for mammals.

Lithology – The study of rocks. Rocks are mixtures of minerals and / or organic / fossil material.

Monophyletic – For a population or taxonomic group, all members of the group are more closely related to each other than to any other population or taxa not a part of the grouping. Moving back in time a monophyletic group must contain all primary ancestors of all members of the group, and all descendents of those ancestors which are primarily derived from those ancestors.

Morphology – The physical attributes of an organism. Which can be seen by the eyes such as under magnification, and parts of which are preserved in fossils. Differences in morphology between different individuals assist in classification of organisms and understanding relationships between organisms.

Mosaic – An individual or population showing a mosaic of traits seen in other different populations, which do not come from a shared ancestor of all the populations contributing mosaic traits. Such traits come about by introgression between populations or by convergent evolution. Mosaic traits are by comparison, to the populations with the same trait, and with populations where the trait is not known. If one population is seen to have a trait, and another does not have the trait but had an ancestor which did have the trait, then that trait is not a mosaic trait it is a plesiomorphic condition.

Natural selection – A process through which organisms which are more relatively fit are more likely to yield larger numbers of fertile offspring than organisms which are less fit. Fitness being in large part determined by the genomes suitability to the life situation of the individual. Over time individual genes can in effect be acted upon by natural selection.

Paleoanthropology – The study of human fossil remains. Physical anthropology is closely related.

Plesiomorphic trait – A trait shared between two groups which was present in a common ancestor of those groups. Or a trait seen in one group, but lost in another group, where the two groups share a common ancestor with the trait.

Polyphyletic – A taxa, group of taxa, or group of populations for which some members of the group are more closely related to populations not included in the group than to some member(s) of the grouping. Whales and fish as a group exclusive of horses is a polyphyletic grouping.

Population – A group of individuals that over time breed within their own group more often than they breed with other groups.

Prognathy – The degree to which the jaws jut forward from the face. Habitually bipedal primates show a tendency to have lower prognathy than quadrupedal primates. African and European humans tend to have higher prognathy than East Asian humans of similar age such as Mata Menge Homo floresiensis. Orangutans have low prognathy relative to African apes.

Purative selection – In more specialized populations parts of the genome favoring the habits of the host can be strongly selected for, with other genes not specialized to the habits of the host selected out. By this process genetic diversity can be reduced within a relatively specialized population.

Race – An informal rank below the level of subspecies. A race should be partially but not fully genetically isolated from all other populations of the same species. Any member of a race should be distinguishable from any member of any other race by some genetic or morphological standards. Populations which gently grade into each other, even if quite distinct at certain points, do not constitute races.

Recent Out of Africa hypothesis (ROoA) – A small single population of humans from Africa between 20 and 70 thousand years ago completely replaced all other people and are the sole ancestors of all people not living in Africa today. This has been abandoned in the last few years and revised with different versions in place now. Generally (revised) ROoA states that in the last 70 to 120 thousand years, one or more populations left Africa and essentially replaced all other people outside of Africa. With some minor introgressions from archaic humans in Eurasia. Some versions castrate the neandertal male ancestrally (we have no male neandertal ancestry), others do not.

Sahul – A continent which appears during cold periods due to sea level drop which includes New Guinea Australia and Tasmania.

Speciation – When two or more populations which are mostly derived from a single species, can no longer usually upon contact freely mate to yield usually fertile offspring.

Species – For mammals, a group of individuals, no matter how diverse morphologically or genetically, which if in contact can freely interbreed to yield normally fertile offspring.

Stratigraphy - The branch of geology concerned with the order and relative position of strata and their relationship to the geological time scale.

The analysis of the order and position of layers of archaeological remains. Older below younger above in the order deposited.

Sunda – With sea level drop an area including much land now submerged SE Asia and the Malay archipelago.

Taphonomy – The study of all the processes that act upon an organism between the time of its death and diagenesis. Taphonomic studies end when no trace of an organism remains to be fossilized.

Taxa – A group of populations seen to form a unit, usually given rank such as variety race subspecies species subgenus. Especially in paleoanthropology there is often disagreement about the taxonomic status so more generalized terms can be used to avoid commitment to one view.

Truncated – Missing below a certain point. Truncated fossil record, the oldest known fossil is younger than the taxa existed for the region / continent.

REFERENCES

1. Adcock et al 2001 "Mitochondrial DNA sequences in ancient Australians: Implications for modern human origins." Proc Natl Acad Sci USA 98(2): 537-542

2. Alroy. John "The Fossil Record of North American Mammals: Evidence for a Paleocene Evolutionary Radiation" Department of Paleobiology Smithsonian Institution

3. Atmoko et al 2004 "Alternative male reproductive strategies: male bimaturism in orangutans" Cambridge University Press Pp 196-207

4. Ayub et al 2015 "The Kalash Genetic Isolate. Ancient Divergence, Drift, and Selection" Am J Hum Genet 96(5): 775-783

5. Berger et al 2015 "Homo naledi, a new species of the genus Homo from the Dinaledi Chamber, South Africa" doi. 10.7554/eLife09560

6. Berger et al 2008 "small-bodied humans from Palau, Micronesia" PLOS ONE 3. 1-11

7. Bohme et al 2019 "A new Miocene ape and locomotion in the ancestor of great apes and humans" Nature https://doi.org/10.1038/s41586-019-1731-0

8. Boothby et al 2015 "Evidence for extensive horizontal gene transfer from the draft genome of a tardigrade" PNAS

9. Brace, C. Loring 1962 "Refocusing on the Neanderthal Problem" American Anthropologist

10. Callaway, Ewen 2017 "Oldest Homo sapiens fossil claim rewrites our species' history" https://www.nature.com/news/oldest-homo-sapiens-fossil-claim-rewrites-our-species-history-1.22114

11. Curnoe et al 2015 "A Hominin Femur with Archaic Affinities from the Late Pleistocene of Southwest China" PLOS ONE https://journals.plos.org/plosone/article?id=10.1371/journal.pone.0143332

12. Fitzpatrick et al 2008 "Small Scattered Fragments Do Not a Dwarf Make:" PLOS ONE https://doi.org/10.137/g/journal.pone0003015

13. Frayer. D. W. 1973 "Gigantopithecus and its relationship to Australopithecus" American Journal of Physical Anthropology 39(3): 413-426

14. Fuss et al 2017 "Potential hominin affinities of Graecopithecus from the Late Miocene of Europe" PLOS One 12 (5)

15. Gibbons, Ann 2017 "World's oldest Homo sapiens fossils found in Morocco"JebelIrhoudhttps://www.sciencemag.org/news/2017/06/world-s-oldest-homo-sapiens-fossils-found-morocco

16. Gierlinski et al 2017 "Possible hominin footprints from the late Miocene (c. 5.7Ma) of Crete?" Proceedings of the Geologists Association 128 (5-6): 697-710

17. Grehan, J. R. and Schwartz J. H. 2011 "Evolution of human-ape relationships remains open for investigation" Journal of Biogeography, 38. 2397-2404

18. Haber et al 2019 "A Rare Deep-Rooting African Y-Chromosomal Haplogroup and its Implications for the Expansion of Modern Humans Out of Africa" Genetics 212(4): 1421-1428

19. Haile-Selassie et al 2016 "The Pliocene hominin diversity conundrum: Do more fossils mean less clarity?" PNAS 113(23) 6364-6371

20. Haslam, Michael 2014 "On the tool use behavior of the bonobo-chimpanzee last common ancestor, and the origins of hominine stone

tool use" American Journal of Primatology https://onlinelibrary.wiley.com/doi/abs/10.1002/ajp.22284

21. Hawks et al 2000 "An Australasian test of the recent African using the WLH-50 calvarium" Journal of Human Evolution 39, 1-22 http://www-personal.umich.edu/~wolpoff/Papers/wlh-50.pdf
22. Huang, Shi 2009 Inverse relationship between genetic diversity and epigenetic complexity" natureprecedings doi: 10.1038/npre.2009.1751.2
23. Ignicco et al 2018 "Earliest known hominin activity in the Philippines by 709 thousand years ago" Nature 557, 233-237
24. Kimbel, William H. and Villmore, Brian 2016 "From Australopithecus to Homo: the transition that wasn't" Philosophical Transactions of the Royal Society B
25. Kulwilm et al 2019 "Ancient admixture from an extinct ape lineage into bonobos" nature ecology & evolution 3, 957-965
26. Leakey et al 1964 "A New Species of the Genus Homo from Olduvai Gorge" Nature 202 (4927):7-9
27. Leakey, Louis 1964 "Homo habilis, Homo erectus and the australopithecines:" Nature 209 1279-1281
28. Martinon-Torres et al 2019 "Further dental analysis of Homo antecessor suggests it was basal to Homo sapiens, Homo neanderthalensis and the Denisovan Hominins" https://www.ucl.ac.uk/human-evolution/news/2019/jan/further-dental-analysis-homo-antecessor-suggests-it-was-basal-homo-sapiens-homo
29. Meyer et al 2016 "Nuclear DNA sequences from the Middle Pleistocene Sima de los Huesos hominins" Nature 531, 504-507
30. Milks et al 2019 "External ballistics of Pleistocene hand-thrown spears: experimental performance data and implications for human evolution" Scientific Reports Article number 820

31. Peyregne et al 2019 "Nuclear DNA from two early Neandertals reveals 80,000 years of genetic continuity in Europe" Science Advances Vol 5, no 6, eaaw 5873

32. Pickford, M. and Senut, B. 2001 "Millenium Ancestor', a 6-million-year-old bipedal hominid from Kenya-Recent discoveries push back human origins by 1.5 million years." South African Journal of Science 97, 22-22.

33. Pickford, Martin and Senut, Brigitte 2001 "The geological and faunal context of Late Miocene hominid remains from Lukeino, Kenya." C. R. Acad. Sci. Paris 332, 145-152

34. Pickford, M. and Senut, B. 2004 "Hominoid teeth with chimpanzee-and-gorilla-like features from the Miocene of Kenya:" Anthropological Science Vol. 113, 95-102

35. Proffitt et al 2016 "Wild monkeys flake stone tools" Nature 539, 85-88

36. Pruetz, Jill D. and Bertolani, Paco 2007 Current Biology https://www.cell.com/current-biology/fulltext/S0960-9822(07)00801-9

37. Reich, David 2018 "Who We are and how We Got Here" book ISBN-10: 110187032X

38. Richmond, Brian G. and Straight David S. 2000 "Evidence that humans evolved from a knuckle-walking ancestor" Nature 404, 382-385

39. Schoch et al 2015 "New insights on the wooden spears from the paleolithic site of Schoningen" Journal of Human Evolution Volume 89, 214-225

40. Simons, Elwyn L. 1963 "Some Fallacies in the Study of Hominid Phylogeny" Science 06 Vol 141, Issues 3584 pp 879-889

41. Skoglund et al 2015 "Genetic evidence or two founding populations of the Americas" Nature 525, 104-108

42. Spassov et al 2012 "A hominid tooth from Bulgaria: The last pre-human hominid of continental Europe." Journal of Human Evolution 62(1)

43. Steele et al 2013 "Comparative Morphology of the Hominin and African Ape Hyoid Bone, a Possible Marker of the Evolution of Speech" Human Biology, 85(5): 639-672

44. Stringer C. B. 2003 "Out of Ethiopia" Nature, 423: 692-4

45. Tierny et al 2017 "Rainfall Regimes of the Green Sahara" Science Advances

46. Van den Bergh et al 2016 "Homo floresiensis-like fossils from the early Middle Pleistocene of Flores" Nature 534, 245-248

47. Venn et al 2014 "Strong male bias drives germline mutation in chimpanzees" Science Vol. 344 Issue 6189pp 1272-1275

48. Washburn, S. L. and Ciochon, R. L. 1974 "Canine Teeth: Notes on Controversies in the Study of Human Evolution"" https://anthrosource.onlinelibrary.wiley.com/doi/pdf/10.1525/aa.1974.76.4.02a00030

49. Weidenreich, Franz 1946 "Apes Giants and Man" ISBN-13: 978-0226881478

50. Welker et al 2019 "Enamel proteome shows that Gigantopithecus was an early diverging pongine" Nature 576, 262-265

51. White et al 1994 "Australopithecus Ramidus, a new species of early hominid from Aramis, Ethiopia." Nature, 375 (6526), 88.

52. White et al 2003 "Pleistocene Homo sapiens from Middle Awash, Ethiopia" Nature, 423: 742-7

53. Wolpoff, Milford H. 1982 "Ramapithecus and Hominid Origins" Current Anthropology 23(5)

54. Wood, Bernard 2002 "Hominid revelations from Chad" Nature 418

55. Wood, Bernard A. and Constantino, Paul J. 2007 "Paranthropus boisei: Fifty Years of Evidence and Analysis" Yearbook of Physical Anthropology 50: 106-132

56. Yuan et al 2017 "Modern human origins: multiregional evolution of autosomes and East Asia origin of Y and mt DNA" bioRxiv pre-print https://www.biorxiv.org/content/10.1101/101410v1

57. Xiu-Jie Wu et al 2019 "Archaic human remains from Hualongdong, China, and Middle Pleistocene human continuity and variation" PNAS 116 (20) 9820-9824

58. Xueping et al 2012 "Human remains from the Pleistocene Holocene transition of southwest China suggest a complex evolutionary history for East Asians" PLOS ONE 7(3): e 31918

ABOUT THE AUTHOR

Born Alan Darwin VanArsdale in 1961 in the Western United States. Has lived in Mexico China Ukraine and Honduras. Awarded AA at Oxnard City College and BS with one year of graduate studies from the College of Creative Studies at the University of California Santa Barbara in creative studies emphasis graduate level biology research in 2001. Previous publication includes the book "Roman Coin Forgery" 2005 a referenced work in forensic numismatics published in Bulgaria. Most recent employment numismatist and store owner at Ogden Utah last 15 years. The author has logged over 1000 days and nights camping in wilderness collecting terrestrial fossil mammals. Recovering more than 30,000 diagnostic fossil vertebrates. Mostly rodents but including many larger mammals such as elephants (Gomphotheres) fossils. With over 300 species and over 50 new species including a published new species of horse. Most fossils which are housed at the Los Angeles County Museum of Natural History, where the author was briefly employed including as a field party leader, the San Bernardino County Museum of Natural History and the Museum at the Texas A & M University. The author taught a paleontology course, adult education, for the Ventura County Parks and Recreation department including a field trip. The author was awarded a one-year stipend for field work before attending the University of California under the supervision of Dr Andre Wyss (vertebrate paleontologist), who was later one of the author's three academic advisors in graduate studies. At UCSB the author was awarded a university presidential fellowship to obtain absolute dates from field samples of volcanoclastic materials and led one UCSB anthropology field party studying lithic source materials of the Chumash tribe.

At about the same time as the Alvarez team discovered diagenetically unaltered microtektites (which were previously unknown except in altered forms), the author discovered diagenetically unaltered microtektites in a terrestrial Miocene horizon in California, which were presented by Alessandro Montanari of the Alvarez team at a University of California Berkeley geophysics conference. The author made a high-resolution paleontological evaluation of the event horizon and determined there had been local rodent extinction (for an otherwise locally common mouse genus Copemys suggesting aridification) after the bolide impact event, with a one species fresh-water gastropod bloom just after the event (a coquina suggesting lack of predation). The author spent much of his childhood in the wilderness of NW Montana and spent the coldest part of the winter of 1981/1982 camping alone in remote mountains of NW Montana. Currently the author is planting and caring for his second remote organic fruit tree orchard, in a high elevation region in Idaho. Past formal studies by the author include vertebrate paleontology with emphasis on Neogene especially Miocene mammals, anthropology, mathematics as a major at Humboldt State University, sedimentary geology, macro-organismal genetics and stratigraphic archaeology. The author administrates the popular face book groups "The into Africa Theory" with the group founder Bruce Fenton who has written a popular book series "The Forgotten Exodus" on past human origins and migrations. And co-administrates the group Denisovan with the group founder, paleontologist Branislav Petkovic and a new administrator during the writing of this book Marc Young current graduate student in paleoanthropology.